# Contributions to Management Science

The series *Contributions to Management Science* contains research publications in all fields of business and management science. These publications are primarily monographs and multiple author works containing new research results, and also feature selected conference-based publications are also considered. The focus of the series lies in presenting the development of latest theoretical and empirical research across different viewpoints.

This book series is indexed in Scopus.

Hatem Masri

Editor

# Africa Case Studies in Operations Research

A Closer Look into Applications
and Algorithms

 Springer

*Editor*
Hatem Masri
College of Business Administration
University of Bahrain
Sakhir, Bahrain

ISSN 1431-1941          ISSN 2197-716X   (electronic)
Contributions to Management Science
ISBN 978-3-031-17010-2          ISBN 978-3-031-17008-9   (eBook)
https://doi.org/10.1007/978-3-031-17008-9

This Springer imprint is published by the registered company Springer Nature Switzerland AG
The registered company address is: Gewerbestrasse 11, 6330 Cham, Switzerland

# Contents

# Traveling Advisor Problem in Occupational Health and Safety Field with a Case Study from Egypt

**Said Ali Hassan, Prachi Agrawal, Talari Ganesh, and Ali Wagdy Mohamed**

**Abstract** Numerous application problems are expressed as nonlinear binary programming models, which make it challenging to address them using precise methods, especially in cases where the dimensions are enormous. A new problem in network optimization known as the Traveling Advisor Problem (TAP) is one of these practical applications. It is defined as an advisor who wants to select a subset of candidate workplaces that comprise the most profitable route within the time constraints of daily working hours to maximize profitability.

To address binary optimization issues, this article suggests a novel binary variant of the Gaining-Sharing knowledge-based optimization technique (GSK). The GSK algorithm is built on the idea of how people learn and impart knowledge throughout their lives. The binary stages of binary junior gaining-sharing stage and binary senior gaining-sharing stage with knowledge factor 1 are the fundamental components of the sharing knowledge-based optimization method (BGSK). These two phases give BGSallow BGSKciently and effectively explore and utilize the search space to address issues in binary space.

A Nonlinear Binary Model is introduced with a detailed real application example; the example is solved using the novel Binary Gaining-Sharing knowledge-based optimization algorithm (BGSK). The obtained optimal solution is better than that provided by the health and safety agency management in utilizing the available time and profitability.

S. A. Hassan
Department of Operations Research and Decision Support, Faculty of Computers and Artificial Intelligence, Cairo University, Giza, Egypt

P. Agrawal · T. Ganesh
Department of Mathematics and Scientific Computing, National Institute of Technology Hamirpur, Hamirpur, Himachal Pradesh, India

A. W. Mohamed (✉)
Operations Research Department, Faculty of Graduate Studies for Statistical Research, Cairo University, Giza, Egypt

Department of Mathematics and Actuarial Science, School of Sciences Engineering, The American University in Cairo, Cairo, Egypt

© The Author(s), under exclusive license to Springer Nature Switzerland AG 2022
H. Masri (ed.), *Africa Case Studies in Operations Research*, Contributions to Management Science, https://doi.org/10.1007/978-3-031-17008-9_1

# 1   Introduction

Workers are exposed to various risks while at work in every country on the globe. They sustain injuries due to slips and falls, overexertion, exposure to dangerous chemicals and materials, and machinery use. Physical risks include noise, vibration, and trip, trip, and fall hazards Salim and others (2017). Chemical risks are potentially harmful substances that could hurt workers, Bhusnure (2018). Biological dangers, while bacteria and viruses like hepatitis, HIV/AIDS, and Legionnaire's disease, S. L. Sacadura et al. (2018).

The Health and Safety Advisor position aims to support and promote safety, health, and well-being across all functional areas of businesses. Supporting the tenets of "Doing Safety Differently," working with personnel and management to ensure compliance with regulations is achieved, and establishing such a culture. Additionally, this position will conduct reviews, audits, and devising strategies for best practices in health and safety Carnell and Nebosh (2017).

A new problem, which we will call the Traveling Advisor Problem (TAP) in network optimization, is defined as an advisor who wants to settle on the foremost profitable route for visiting several candidate workplaces associated with a corresponding profit. He begins from the Occupational Health and Safety Company Headquarter (HQ) and next wants to go to each workplace exactly once within the day working hours. The target of the TAP is to work out the route with maximum profitability for the company. This definition is somewhat resembling that of the Traveling Salesman Problem (TSP) and its variations with some basic indicated differences.

Section 2 is devoted to demonstrating the importance of the Occupational Health and Safety and shining up the reality that workplace injuries and illnesses can be prevented or get smaller if management keeps health and safety regulations at the forefront for every job operation. This section reveals also the role and advices offered by the Occupational and Health Advisor to organizations and workplaces.

Section 3 gives a brief review of the TSP and its variations like the problem of Traveling Salesman with Priority Prizes, Deliveries and Collections, Pickup and Delivery, Backhauls, Multiple depots, Specifications of salesmen number, Problems with time windows and that with fixed charges. It deals also with problems like Multiple Traveling Salesman, Generalized Traveling Salesman, Generalized Vehicle Routing and Traveling Repairman one. The main differences between TAP and the TSP are illustrated.

The mathematical model of a TAP with Single Advisor is designed in Sect. 4 including the definition of problem variables, constraints, and the objective function. The proposed model is a Nonlinear Binary Model with a dimension depending on the number of candidate workplaces, the steps of the solution procedure are also explained.

A real practical example is explained in Sect. 5; its complete mathematical model is formulated including the definition of the binary decision variables, constraints comprising: Positions, Workplaces, Sequencing Route Positions, Working Hours / Day, and then the objective function.

In Sect. 6, a novel Binary version of a recently developed gaining-sharing knowledge-based optimization algorithm (GSK) is introduced to solve the TAP. GSK cannot solve the problem with binary space; therefore, Binary-Gaining-Sharing knowledge-based optimization algorithm (BGSK) is proposed with two new binary junior and senior stages. These stages allow BGSK to explore and exploit the search space of the problem efficiently.

Section 7 represents the experimental results of TAP obtained by BGSK, and Sect. 8 summarizes the conclusions and the suggested points for future researches.

## 2   Occupational Health and Safety Advisory Problem

The importance of the Occupational Health and Safety appears clearly from the International Day of Mourning held every year, where International Labor Organizations and a huge number of people are collected around the world on April 28th to keep an eye on the workers' safety. The Day of Mourning has twofold—to remind and respect, those who lost lives or wounded during the work and to continue the promise to hinder more deaths, wounding and sickness, by enhancing safety and health in the place of work.

The Association of Workers and the Compensations Boards of Canada state that approximately one thousand workers die, and about 250,000 workers suffer work-related injuries or diseases every year. Every year in Canada, more than 48,000 young workers under the age of 25 are injured seriously enough to require time off from work.

The reality is that workplace injuries and illnesses can be prevented or get smaller if we keep health and safety at the forefront every day. Occupational Health and Safety (OH&S) programs have been developed to protect workers and employers from incidents and occupational diseases related to the workplace. An OH&S program also outlines the roles and responsibilities of employers, supervisors, and workers within the workplace.

An OH&S program is a procedure for handling safety and health matters in the place of working. Its benefits extend to employers, supervisors, and workers including those in safer and healthier work environment, lower medical care costs, lower insurance costs, increased work productivity and efficiency, increased profits and fewer work-related tragedies.

OH&S Acts and Regulations are laws that govern workplaces. They outline the rights and responsibilities of the employer, the worker, and the supervisor to ensure working environments are healthy and safe, Coady et al. (2015). Egyptian Law regulations in this field of study are enrolled under the International Labor Organization and the World Health Organization (WHO), Awad and Nour El-Din (2018). The International Labor Organization website (2020) and the Egyptian labor code "Law 12/2003" dedicate a specific section (Book V) to OH&S and confirmation of the suitability of the environment in workplace. It is augmented with Ministerial instructions that detail specifically technical provisions relating to notifications and procedures in the case of work-related diseases, fatalities, injuries, and accidents; specifying precautions and circumstances vital for the OH&S assistance at work; defining OH&S committees and services; OH&S training organizations and other laws and regulations indirectly relate to OH&S.

It is the responsibility of the advising agency to know what circumstances are in the place of work affecting negatively health and safety of employees. Risk estimation helps defining what is required to protect people. They are scheduled and carried out regularly by specialized agencies and local assigned groups. The check-up concentrates on risks at the workplace and those that can happen due to the introduction of new operations, tools, constituent or the revision of existing procedures. These inspections help distinguishing and preventing risk, Sika Group (2020). Working as a safety and health advisor comprises helping employees by assuring that hazards are precisely controlled. Health and safety advisors make sure that people comply with all rules of safety and health, and that work environments are not the source of illness, wounding, or dying, Parker (2017).

Advisors accomplish their work by inspecting accidents, checking buildings and equipment, advising people and through implementing the regulations. They keep an eye on workplaces to ensure they are safe for employees to work at. They visit properties such as education institutions, shops, stores, quarries, manufacturing firms, farms, and construction sites to ensure that work is accurately performed in compliance with regulations, Marriott and Schmidt-McCleave (2018). Other responsibilities of the job include: preparing safety programs and policies, investigating accidents and complaints, measuring vibrations, heat and noise and taking necessary photos and samples, ensuring workers use protective equipment, predicting possible hazards drawn from experience, providing advices on health and safety, recording violations of regulations, gathering illegal evidence, follow up evidence in court, writing remarks and reports, providing training and educational support to employers Manu et al. (2019).

Working hours are typically 8, Sunday to Thursday. In cases of serious incidents, work will include evening and weekend. Job-sharing, part-time work, and career breaks are quite popular, Boyle (2015).

The proposed TAP in network optimization is defined for an advisor who wants to choose the most profitable route for visiting several candidate workplaces associated with specific profit for each. He begins in the Occupational Health and Safety Agency Head-quarter and next want to visit some chosen workplaces, each one exactly once within the day working hours. The objective of the TAP is to determine the precisely chosen route with maximum profitability for the safety agency.

This definition is somewhat like that of the TSP and its variations. This proximity will be useful when creating the mathematical model for the new proposed problem TAP.

## 3 Traveling Advisor Problem Vs. Traveling Salesman Problem

### 3.1 The Traveling Salesman Problem: An Overview

The TSP is one of the issues that is overly taken into account while analyzing networks because it has numerous practical applications, according to Applegate et al. (2006). TSP can be summed up as a salesman traveling to several locations; he starts in his hometown and wishes to visit each location on a list only once. He finally goes back to his hometown. He aims to minimize the overall distance traveled by doing this. The issue seems quite straightforward, but finding a solution is far more challenging. Mathematicians have been attempting to solve it for almost a century. The issue has $(n - 1)!/2$ potential solutions, which are limited but difficult to count, Gleixner (2014).

According to Droste (2017), there are many various types of tours, thus one might not assume to solve the issue by merely figuring out how long each one might be. There are more than 3 billion conceivable excursions for just 14 locations, thus this is only feasible in very specific circumstances. Therefore, to resolve these issues, we require appropriate algorithms.

TSP is one of the issues in combinatorial optimization that has received the most attention, and it has been shown to be non-deterministic polynomial-time hard (NP-hard). TSP was already recognized as a concern in the nineteenth century. Mathematicians started to take an interest in it in the 1930s. Since then, more advanced algorithms have been created. The largest solved instance in 1954 included 49 US cities. At the time, this was a significant accomplishment. The quickest route through all 24,978 cities in Sweden was discovered in 2004. A computer chip with 85,900 places is the current record. Appelgate and co. were able to solve it (2009). The papers address a variety of TSP variations, and a large number of these variations frequently have their roots in foundational programming concepts Sarubbi and Luna (2007).

The TSP with Priority Prizes is considered as an expansion of the classical TSP, where the traveling salesman visits all clients and the order of the client's visits are considered in the desired goal. The prize is awarded when an indicated client is reached in a specific chosen rank. The cost of traveling is considered when moving from any client to another in a chosen route. The objective of the TSPPP is to maximize all the customer visits profit taking the involved costs and prizes in the route, Pureza et al. (2018).

Baldacci et al. (2003) investigate a formulation with integer variables for an expansion of the TSP by adding mixed deliveries and collections, which is the TSP with Pickup and Delivery or TSP with Deliveries and Collections (TSPDC), where the served customer set is divided to 2 subgroups: collection and distribution. A vehicle of a specific load is located in a centralized warehouse and utilized to provide delivery clients and gather matters from collection clients. In the TSPDC, the matters to be distributed are not of the same type as that to be collected. A salesman departs from the warehouse having a total matter as that required from distribution clients and goes back to the warehouse having a total matter as that of the collection clients. The TSPDC task is to define a tour beginning and terminating at the depot, visiting each client only once, while minimizing the tour length. The load of the vehicle should not surpass its maximum tonnage. When all distribution clients must be visited before the collection clients, the case is defined to be the TSP with Backhauls (TSPB), Gendreau et al. (1996) and Aramgiatisiris (2004).

Mosheiov (1994) suggested a formulation of TSPDC, in plus he introduced heuristic models supported with extended formulations of TSP. Anily and Mosheiov (1994) described a novel approach derived from the algorithm of Shortest Spanning Trees. Gendreau et al. (1999) suggested two approaches for TSPDC. The generalization of the TSPDC related to the VRP, whereas multiple vehicles are considered, is proposed, Halse (1992), he suggested a heuristic approach and a mathematical algorithm relying on a Lagrangian relaxation. Dumitrescu et al. (2010) modeled a mixed-integer programming model and its polyhedral framework for the TSPPD. The dimension of the model polytope is specified, and various adequate inequalities are proposed. Heuristics for the TSP with backhauls (TSPB) were introduced by Gendreau et al. (1996).

Given an undirected graph whose nodes are partitioned into several subsets (clusters), the Generalized Traveling Salesman Problem (GTSP) is searching a tour with minimum cost which has only one node for every subset. The TSP is a specific situation of GTSP in which every subset includes only one node. Pop (2007) described six distinct integer programming formulations and established relationships between the polytopes related to the linear relaxations of the GTSP.

Kara and Bektas (2003) presented a mixed-integer programming formulation of the Generalized Vehicle Routing Problem (GVRP), it is similar to GTSP, which is a prolongation of the Vehicle Routing Problem (VRP), but the customers are collected into several node-sets. The performance of the computation of the suggested formulations is determined by commercial packages applied to approved standard problems.

The Multiple Traveling Salesman Problem ($m$TSP) is a general concept obtained by inference from the Problem of Traveling Salesman where more than one salesman is available. Considering a group of cities, a storehouse in which the salesmen will start and finish their routes, and a cost measure. Each city is visited only once by one of the salesmen, the problem aim is distinguishing a route for every salesman, which minimizes the overall cost of the routes. Bektas (2006) names some variations of the $m$TSP:

– Multiple depots: These problems possess a number of storehouses greater than one, a number of salesmen belong to each storehouse. In the problem with one destination, each salesman goes back to the specific storehouse he begins. For the not necessarily storehouse itself problem, a salesman is not forced to go back to the storehouse itself at which he began, but the equal number of salesmen should go back as commenced at a storehouse. This problem is applied in robots with land and air vehicles, Oberlin et al. (2009).

– Specifications on the number of salesmen: Salesmen's number can be constant, or it can be determined by the solution but is upper bounded.
– Fixed charges: In the version of not-fixed number of salesmen in the $m$TSP, there is a fixed charge for operating a salesman. In the version of fixed charges, the overall cost calculates the charges of activated salesmen and the traveling expenses.
– Time windows: There is a variant of the $m$TSP with time windows ($m$TSPTW) as with the TSP and the VRP. A time slot is related to every node in which it must be visited by a salesman. The $m$TSPTW is applied in routing school-bus and in scheduling air travels.

Demiral (2021) constructed a model that investigates the characteristics of TSP in two routes of a Double Traveling Salesman Problem ($d$ TSP). It is a variant of $m$-TSP with two separate salesmen operate at the same time, they begin and return either in one or two depots. They used Simulated Annealing to optimize the tours of salespersons.

The Traveling Repairman Problem (TRP): It is known also as a traveling deliveryman, cumulative traveling salesman and minimum latency problem. It is a specific kind of routing problem, Silva et al. (2012). In this kind, customer latency is determined from the tour start of the route until the completion of the client's servicing. The $m$TRP is an extension of TRP since it finds $a$ tour for each repairman, each one begins from the store, visiting some determined customers that minimize the total latency, Onder et al. (2017).

Algorithms for solving TSP and $m$TSP are divided into exact and approximate (heuristic) algorithms. An exact algorithm guarantees to find the shortest tour. A heuristic algorithm will find a good tour, but it is not guaranteed that this will be the best tour. The advantage of a heuristic algorithm is the shorter running time which makes it more suitable for large instances. Integer Programming is one of the widely used methods to formulate and then find optimal solutions for the TSP.

Orman and Williams (2006) survey eight models of the Asymmetric TSP as Integer Programs (IP). They assort them as: time staged, flow-based, sequential, and conventional. Conventional. Time Staged: first and second Stage Dependent, Fox et al. (1980), and third Stage Dependent, Vajda (1961). Sequential Formulation, Miller et al. (1960) and Sawik (2016). Flow-Based Forms: Single-Commodity, Gavish and Graves (1978), Two-Commodity, Finke et al. (1983) and Multi-Commodity, Wong (1980) and Claus (1984) and the Conventional Formulation, Dantzig et al. (1954).

## 3.2 Main Differences Between TAP and TSP

TAP differs from the famous and well-known TSP in the following main points:

1. In the TSP, the time is open till completing visiting of all customers, while in the TAP, the available time is limited by the day working hours.

**Table 1** Main differences between TAP and TSP

| Item | TAP | TSP |
|---|---|---|
| Order of picking-up items or visiting places | Effectively considered | Effectively considered |
| Completion time for problem | Limited | Open |
| Served customers | Workplaces | Customers |
| Number of served customers | Some or all | All |
| Selection criteria | Traveling and advising times | Traveling times |
| Objective function | Maximize profits | Minimize total traveling time |

2. In the TSP, the salesman will visit all the customers, while in the TAP, the Advisor will determine a route containing some or all the workplaces, which optimizes the problem objective function within the available limited day time.
3. In the TSP, no time is consumed in customer places, or it is immaterial, while in the TAP, the advising time at a workplace is a basic factor in the day time limit constraint and hence, it directly affects the choice of the optimum solution of the problem.
4. In the TSP, the objective is to complete the route while minimizing the total traveling time, while in the TAP the objective function is to maximize the total profit in the chosen route.

The main basic differences between TCP and both TSP and KP can be summed up as indicated in Table 1:

## 4 Traveling Advisor Problem with Single Advisor

The new TAP is defined on a graph $G$ with a set of $n$ nodes V representing the workplaces (customers) and an additional node denotes the Health and Safety Agency headquarter (HQ) where the advisor starts the job, and a set of arcs representing the traveling times between two distinct workplaces, Pinter (2014). The time of inspection in a workplace and the traveling time between two workplaces are specified.

While the advisor should start his route at the Headquarter (HQ), he ends his route at the last visited workplace.

A Traveling Advisor Problem with single advisor or (a group of advisors working as one team in any workplace) TAP is defined as follows:

- Each workplace is visited only once by the advisor.
- The advisor route starts at the headquarter and terminates at the latest visited workplace.

- The objective to be achieved is the maximization of the overall profit of a route chosen by the advisor under the time limit constraint represented by the working hours/day.

Mathematical Model:

Decision Variables:

Let:

$$x_i^m = \begin{cases} 1, & \text{if workplace } i \text{ is visited by the Advisor on position } m \text{ of his route, } i \text{ and } m = 1, 2, \ldots, n. \\ 0, & \text{otherwise.} \end{cases}$$

Constraints:

1. Positions Constraints:

Each position $m$ in the Advisor route has at most one workplace:

$$\sum_{i=1}^{n} x_i^m \leq 1, m = 1, 2, \ldots, n. \tag{1}$$

2. Workplaces Constraints:

Each workplace $i$ can be in one position of the Advisor route or not visited:

$$\sum_{m=1}^{n} x_i^m \leq 1, i = 1, 2, \ldots, n. \tag{2}$$

3. Consecutive Positions Constraints:

A position $(m + 1)$ cannot exist in the Advisor tour unless the preceding position $m$ exists, this is achieved by the following set of constraints:

$$\sum_{i=1}^{n} x_i^{m+1} \leq \sum_{i=1}^{n} x_i^m, m = 1, 2, \ldots, n - 1 \tag{3}$$

$$\text{If } \sum_{i=1}^{n} x_i^{m+1} = 1, \text{then } \sum_{i=1}^{n} x_i^m = 1, m = 1, 2, \ldots, n - 1,$$

$$\text{If } \sum_{i=1}^{n} x_i^{m+1} = 0, \text{then there is no restriction on the value of } \sum_{i=1}^{n} x_i^m, m = 1, 2, \ldots, n - 1.$$

4. Working Hours/Day Constraints:

The total time spent by the Advisor in traveling and site inspection should be within the maximum working hours/day $= T = 8$ hours.

$$\sum_{i=1}^{n} t_{0i} x_i^1 + \sum_{m=1}^{n} \sum_{i=1}^{n} \left( t_i x_i^m \right) + \sum_{i=1}^{n} \sum_{\substack{j=1 \\ j \neq i}}^{n} t_{ij} \cdot \left( \sum_{m=1}^{n-1} x_i^m \cdot x_j^{m+1} \right) \leq T. \qquad (4)$$

Where:

$t_{0i}$ = Transportation time between the headquarter and workplace $i$, $i = 1, 2, \ldots,$ $n$,

$t_{ij}$ = Transportation time between the two adjacent workplaces $i$ and $j$, $i, j = 1,$ $2, \ldots, n$,

$t_i$ = Inspection time for workplace $i$, $i = 1, 2, \ldots, n$.

This is a quadratic inequality in two variables, the first part is for traveling from the headquarter to the first position in the route, the second part is the inspection times at workplaces and the third part is the traveling time between different positions in the route.

5. Binary Constraints:

All the decision variables are 0–1.

$$x_i^m = 0 \text{ or } 1, i, m = 1, 2, \ldots, n. \qquad (5)$$

6. The Objective Function:

It is formulated for maximizing the total profits of the Occupational Health and Safety agency gained by visiting the workplaces during the working day time limit:

$$\text{Max } z = \sum_{m=1}^{n} \sum_{i=1}^{n} p_i x_i^m \qquad (6)$$

Where: $p_i$ = profit of visiting workplace $i$, $i = 1, 2, \ldots, n$.

So, we have a suggested model that contains $(n^2)$ binary variables and $(3n)$ constraints. For example, with a problem of $n = 10$ workplaces, the number of binary decision variables = 100 and the number of constraints = 30.

The optimum solution will produce two distinct situations:

1. If $\sum_{m=1}^{n} x_i^m = 1, i = 1, 2, \ldots, n$, then all the $n$-workplaces are visited by the Advisor in one working day and the problem is completed.

2. If $\sum_{m=1}^{n} x_i^m = 0$ for any $i$, then the corresponding workplace $i$ is not visited by the Advisor in the first working day. In this case, it is needed to eliminate the visited workplaces, adding one more day and repeat the procedure for another working day.

The steps of the solution procedure are shown in Fig. 1.

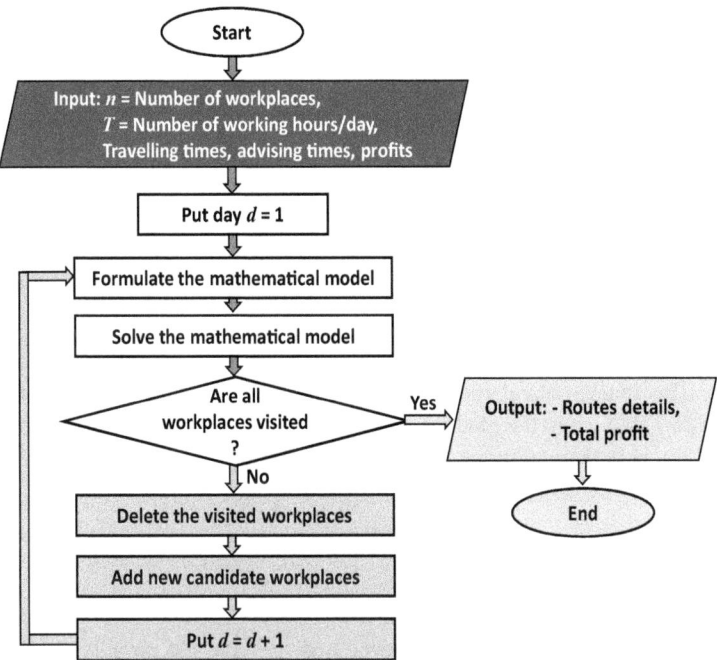

**Fig. 1** Steps of the solution procedure of the TAP

# 5 A Real Application Example

Changes in the environments, technology, and economy create new safety requirements that originate new workplace hazards. Making employees safer at workplaces should be a culture where safety is a common responsibility and a vital value. Whether the firm is a beginning to decide in safety regulations, or it embeds safety rules as a culture for a long time, it needs a comprehensive safety and health advising for management and employees. Companies are continually looking for new ways to embed safety into their work culture, and for one organization, it is one-to-one safety conversations. As part of a broader company-wide safety program, the one-to-one safety conversations aim to reinforce safe behavior and give employees the confidence to engage with others on safety topics. But after running these conversation programs for some time, the organization wanted to know if they were still effective in preventing injury.

The occupational health and safety advisor will provide experienced advices for the health and safety program developers, prepare safety management systems, safety strategies, safety processes, incident screening, planning of emergency response, and handling of post-injury activities.

In this real application case study, an occupational health and safety agency in Cairo, Egypt, has only one advisory group consisting of 3 specialists working as one team in OH&S advices. There are 5 candidate workplaces defined by serial numbers

| $p_i$ (\$) × ($10^2$) | $t_i$ | i \ j | 1 | 2 | 3 | 4 | 5 |
|---|---|---|---|---|---|---|---|
| | | 0 | 0.25 | 0.75 | 0.5 | 1 | 1.25 |
| 3.1 | 3 | 1 | | 0.5 | 0.75 | 1.25 | 1.5 |
| 4.7 | 2 | 2 | | | 1 | 2 | 0.5 |
| 5.5 | 4 | 3 | | | | 1 | 0.75 |
| 4.2 | 5 | 4 | | | | | 1.25 |
| 3.5 | 1.5 | 5 | | | | | |

**Fig. 2** Data for the case study example

(1, 2, . . ., 5), they are located at different positions in Great Cairo Governorate with the data given in Fig. 2 , where numbers inside the cells $(i, j)$ represent the traveling times $t_{ij}$. The mathematical model for the problem is formulated by substituting values from Table 1 into the mathematical formulas (1) to (6). The proposed solution methodology is explained in the next sections.

## 6  Proposed Methodology

Metaheuristic algorithms have been developed to solve the complex optimization problem with continuous variables El-Qulity and Mohamed (2016a, 2016b), El-Qulity et al. (2015, 2016), Mohamed and Mohamed (2019), Mohamed et al. (2011, 2019a, 2019b), Agrawal et al. (2021, 2022), Song et al. (2021). Mohamed et al. (2020) recently proposed a novel Gaining-Sharing Knowledge-based optimization algorithm (GSK), which is based on the ideology of acquiring knowledge and share it with others throughout their lifetime. The original GSK solves optimization problems over continuous space, but it cannot solve the problem with binary space. So, a new variant of GSK is introduced to solve the proposed TAP. A novel Binary Gaining-Sharing knowledge-based optimization algorithms (BGSK) is proposed over discrete binary space with new binary junior and senior gaining and sharing stages.

On the other hand, there are many constraint handling techniques in the literature Deb (2000), Cello (2002), Muangkote et al. (2019). In this paper, the augmented Lagrangian method is used to handle the constraints, in which a constrained optimization problem is converted into an unconstrained optimization problem Long et al. (2013), Bahreininejad (2019). The proposed methodology is described below:

### 6.1  Gaining-Sharing Knowledge-Based Optimization Algorithm (GSK)

A constrained optimization problem can be formulated mathematically as

$$Min\, f(X); X = [x_1, x_2, \ldots, x_{Dim}]$$

$$s.to.g_i(X) \leq 0; i = 1,2, \ldots, m$$

$$X \in [\alpha_p, \beta_p]; p = 1,2, \ldots, Dim$$

Where, $f$ denotes the objective function; $X = [x_1, x_2, \ldots, x_{Dim}]$ are the decision variables; $g_i(X)$ are the inequality constraints and $\alpha_p$, $\beta_p$ are the lower and upper bounds of decision variables, respectively, and Dim represents the dimension of individuals. If the problem is in maximization form, then consider minimization $= -$ maximization.

The junior and senior gaining and sharing stages make up the human-based algorithm GSK. Everyone learns new things and expresses their opinions to others. People in the early stages learn things from their limited networks, which may include family, relatives, neighbors, etc. They are curious about other people and want to express their thoughts with them. These can lack the knowledge necessary to classify persons. Similar to this, people in their middle or later years enrich their knowledge by talking with friends, co-workers, social media friends, etc. and sharing their opinions with the most appropriate person to do so. These individuals can judge and classify others based on their experiences (good or bad). The following steps can be used to mathematically express the aforesaid procedure:

Step 1: To obtain the starting solution of the optimization problem, the initial population must be obtained. The initial population is created randomly within the boundary constraints as:

$$x_{tp}^0 = \alpha_p + rand_p\left(\beta_p - \alpha_p\right) \tag{7}$$

Where: t is for the number of populations; $rand_p$ denotes uniformly distributed random number between $0$ and $1$.

Step 2: At the beginning, the dimensions of the junior and senior stage should be computed through the following formula

$$Dim_J = Dim \times \left(\frac{Gen^{max} - G}{Gen^{max}}\right)^k \tag{8}$$

$$Dim_S = Dim - Dim_J \tag{9}$$

where, $k$ ($>0$) denotes the knowledge rate, that controls the experience rate. $Dim_J$ and $Dim_S$ represent the dimension for the junior and senior stage, respectively. $Gen^{max}$ is the maximum number of generations, and G denotes the generation number.

Step 3: Junior gaining-sharing knowledge stage: In this stage, the early-aged people gain knowledge from their small networks and share their views with the other people who may or may not belong to their group. Thus, individuals are updated through as:

```
for t=1:NP
    for p=1:Dim
        if rand≤ k_r (knowledge ratio)
            if f(x_t) > f(x_r)
```

$$x_{tp}^{new} = \left(x_t + k_f * \left((x_{t-1} - x_{t+1}) + (x_r - x_t)\right)\right)$$

```
            else
```

$$x_{tp}^{new} = \left(x_t + k_f * \left((x_{t-1} - x_{t+1}) + (x_t - x_r)\right)\right)$$

```
            end (if)
            else    x_{tp}^{new} = x_{tp}^{old}
        end (if)
    end (for p)
end (for t)
```

**Fig. 3**  Pseudo-code for Junior gaining-sharing knowledge stage

i. According to the objective function values, the individuals are arranged in ascending order. For every $x_t$ ($t = 1, 2, \ldots, NP$), select the nearest best ($x_{t-1}$) and worst ($x_{t+1}$) to gain knowledge, also choose randomly ($x_r$) to share knowledge. Therefore, to update the individuals, the pseudo-code is presented in Fig. 3.

Where: $k_f(>0)$ is the knowledge factor.

Step 4: Senior gaining-sharing knowledge stage: This stage comprises the impact and effect of other people (good or bad) on the individual. The updated individual can be determined as follows:

i. The individuals are classified into three categories (best, middle, and worst) after sorting individuals into ascending order (based on the objective function values).

Best individual= 100 $p\%$ ($x_{best}$), middle individual= $Dim - 2 * 100p\%$ ($x_{middle}$), worst individual= 100 $p\%$ ($x_{worst}$).

For every individual $x_t$, choose two random vectors of the top and bottom 100 $p$ % individual for gaining part and the third one (middle individual) is chosen for the sharing part. Therefore, the new individual is updated through the following pseudo-code presented in Fig. 4:

Where: $p \in [0, 1]$ is the percentage of best and worst classes.

## 6.2 Binary Gaining-Sharing Knowledge-Based Optimization Algorithm (BGSK)

To solve problems in discrete binary space, a novel Binary Gaining-Sharing knowledge-based optimization algorithm (BGSK) proposed. In BGSK, the new

```
for t=1:NP
    for p=1:Dim
        if rand≤ k_r (knowledge ratio)
            if f(x_t) > f(x_middle)
```
$$x_{tp}^{new} = \left(x_t + k_f * \left((x_{best} - x_{worst}) + (x_{middle} - x_t)\right)\right)$$
```
            else
```
$$x_{tp}^{new} = \left(x_t + k_f * \left((x_{best} - x_{worst}) + (x_t - x_{middle})\right)\right)$$
```
            end (if)
            else   x_tp^new = x_tp^old
        end (if)
    end (for p)
end (for t)
```

**Fig. 4** Pseudo-code for Senior gaining-sharing knowledge stage

initialization and the working mechanism of both stages (junior and senior gaining-sharing stages) are introduced over binary space, and the remaining algorithms remain the same as the previous one. The working mechanism of BGSK is presented in the following subsections:

### 6.2.1 Binary Initialization

The initial population is obtained in GSK using Eq. (18) and it must be updated using the following equation for binary population:

$$x_{tp}^0 = round(rand(0,\ 1))\tag{10}$$

Where: the round operator is used to convert the decimal number into the nearest binary number.

### 6.2.2 Binary Junior Gaining and Sharing Stage

The binary junior gaining and sharing stage is based on the original GSK with $k_f = 1$. The individuals are updated in original GSK using the pseudo-code (Fig. 3), which contains two cases. These two cases are defined for binary stage as follows:

Case 1. When $f(x_r) < f(x_t)$: There are three different vectors $(x_{t-1}, x_{t+1}, x_r)$, which can take only two values (0 and 1). Therefore, a total of $2^3$ combinations are possible, which are listed in Table 2. Furthermore, these eight combinations can be categorized into two different subcases ((a) and (b)), and each subcase has four combinations. The results of each possible combination are presented in Fig. 2.

Subcase (a): If $x_{t-1}$ is equal to $x_{t+1}$, the result is equal to $x_r$.

**Table 2** Results of the binary junior gaining and sharing stage of Case 1 with $k_f = 1$

|              | $x_{t-1}$ | $x_{t+1}$ | $x_r$ | Results | Modified Results |
|--------------|-----------|-----------|-------|---------|------------------|
| Subcase (a)  | 0         | 0         | 0     | 0       | 0                |
|              | 0         | 0         | 1     | 1       | 1                |
|              | 1         | 1         | 0     | 0       | 0                |
|              | 1         | 1         | 1     | 1       | 1                |
| Subcase (b)  | 1         | 0         | 0     | 1       | 1                |
|              | 1         | 0         | 1     | 2       | 1                |
|              | 0         | 1         | 0     | -1      | 0                |
|              | 0         | 1         | 1     | 0       | 0                |

**Table 3** Results of the binary junior gaining and sharing stage of Case 2 with $k_f = 1$

|              | $x_{t-1}$ | $x_t$ | $x_{t+1}$ | $x_r$ | Results | Modified Results |
|--------------|-----------|-------|-----------|-------|---------|------------------|
| Subcase (c)  | 1         | 1     | 0         | 0     | 3       | 1                |
|              | 1         | 0     | 0         | 0     | 1       | 1                |
|              | 0         | 1     | 1         | 1     | 0       | 0                |
|              | 0         | 0     | 1         | 1     | -2      | 0                |
| Subcase (d)  | 0         | 0     | 0         | 0     | 0       | 0                |
|              | 0         | 1     | 0         | 0     | 2       | 1                |
|              | 0         | 0     | 1         | 0     | -1      | 0                |
|              | 0         | 0     | 0         | 1     | -1      | 0                |
|              | 1         | 0     | 1         | 0     | 0       | 0                |
|              | 1         | 0     | 0         | 1     | 0       | 0                |
|              | 0         | 1     | 1         | 0     | 1       | 1                |
|              | 0         | 1     | 0         | 1     | 1       | 1                |
|              | 1         | 1     | 1         | 0     | 2       | 1                |
|              | 1         | 0     | 1         | 1     | -1      | 0                |
|              | 1         | 1     | 0         | 1     | 2       | 1                |
|              | 1         | 1     | 1         | 1     | 1       | 1                |

Subcase (b): When $x_{t-1}$ is not equal to $x_{t+1}$, then the result is the same as $x_{t-1}$ by taking $-1$ as 0 and 2 as 1.

The mathematical formulation of Case 1 is as follows:

$$x_{tp}^{new} = \begin{cases} x_r; \text{if } x_{t-1} = x_{t+1} \\ x_{t-1}; \text{if } x_{t-1} \neq x_{t+1} \end{cases}$$

Case 2. When $f(x_r) \geq f(x_t)$: There are four different vectors $(x_{t-1}, x_t, x_{t+1}, x_r)$, that consider only two values (0 and 1). Thus, there are $2^4$ possible combinations that are presented in Table 3. Moreover, the 16 combinations can be divided into two subcases ((c) and (d)) in which (c) and (d) has four and twelve combinations, respectively.

Subcase (c): If $x_{t-1}$ is not equal to $x_{t+1}$, but $x_{t+1}$ is equal to $x_r$, the result is equal to $x_{t-1}$.

Subcase (d): If any of the condition arise $x_{t-1} = x_{t+1} \neq x_r$ or $x_{t-1} \neq x_{t+1} \neq x_r$ or $x_{t-1} = x_{t+1} = x_r$, the result is equal to $x_t$ by considering $-1$ and $-2$ as 0, and 2 and 3 as 1.

The mathematical formulation of Case 2 is as

$$x_{tp}^{new} = \begin{cases} x_{t-1}; if \ x_{t-1} \neq x_{t+1} = x_r \\ x_t; \text{Otherwise} \end{cases}$$

### 6.2.3   Binary Senior Gaining and Sharing Stage

The working mechanism of binary senior gaining and sharing stage is the same as the binary junior gaining and sharing stage with value of $k_f = 1$. The individuals are updated in the original senior gaining-sharing stage using pseudo-code (Fig. 4) that contain two cases. The two cases further modified for binary senior gaining-sharing stage in the following manner:

Case 1. $f(x_{middle}) < f(x_t)$: It contains three different vectors ($x_{best}$, $x_{middle}$, $x_{worst}$), and they can assume only binary values (0 and 1), thus total eight combinations are possible to update the individuals. These total eight combinations can be classified into two subcases [(a) and (b)] and each subcase contains only four different combinations. The obtained results of this case are presented in Table 4.

Subcase (a): If $x_{best}$ is equal to $x_{worst}$ then the obtained results are equal to $x_{middle}$.

Subcase (b): On the other hand, if $x_{best}$ is not equal to $x_{worst}$ then the results are equal to $x_{best}$ with assuming $-1$ or 2 equivalent to their nearest binary value (0 and 1 respectively).

Case 1 can be mathematically formulated in the following way:

**Table 4**   Results of binary senior gaining and sharing stage of Case 1 with $k_f = 1$

|             | $x_{best}$ | $x_{worst}$ | $x_{middle}$ | Results | Modified Results |
|-------------|------------|-------------|--------------|---------|------------------|
| Subcase (a) | 0          | 0           | 0            | 0       | 0                |
|             | 0          | 0           | 1            | 1       | 1                |
|             | 1          | 1           | 0            | 0       | 0                |
|             | 1          | 1           | 1            | 1       | 1                |
| Subcase (b) | 1          | 0           | 0            | 1       | 1                |
|             | 1          | 0           | 1            | 2       | 1                |
|             | 0          | 1           | 0            | $-1$    | 0                |
|             | 0          | 1           | 1            | 0       | 0                |

**Table 5** Results of binary senior gaining and sharing stage of Case 2 with $k_f = 1$

|  | $x_{best}$ | $x_t$ | $x_{worst}$ | $x_{middle}$ | Results | Modified Results |
|---|---|---|---|---|---|---|
| Subcase (c) | 1 | 1 | 0 | 0 | 3 | 1 |
|  | 1 | 0 | 0 | 0 | 1 | 1 |
|  | 0 | 1 | 1 | 1 | 0 | 0 |
|  | 0 | 0 | 1 | 1 | −2 | 0 |
| Subcase (d) | 0 | 0 | 0 | 0 | 0 | 0 |
|  | 0 | 1 | 0 | 0 | 2 | 1 |
|  | 0 | 0 | 1 | 0 | −1 | 0 |
|  | 0 | 0 | 0 | 1 | −1 | 0 |
|  | 1 | 0 | 1 | 0 | 0 | 0 |
|  | 1 | 0 | 0 | 1 | 0 | 0 |
|  | 0 | 1 | 1 | 0 | 1 | 1 |
|  | 0 | 1 | 0 | 1 | 1 | 1 |
|  | 1 | 1 | 1 | 0 | 2 | 1 |
|  | 1 | 0 | 1 | 1 | −1 | 0 |
|  | 1 | 1 | 0 | 1 | 2 | 1 |
|  | 1 | 1 | 1 | 1 | 1 | 1 |

$$x_{tp}^{new} = \begin{cases} x_{middle}; \text{if } x_{best} = x_{worst} \\ x_{best}; \text{if } x_{best} \neq x_{worst} \end{cases}$$

Case 2. $f(x_{middle}) > f(x_t)$: It consists of four different binary vectors ($x_{best}$, $x_{middle}$, $x_{worst}$, $x_t$), and with the values of each vector, a total of sixteen combinations are presented. The sixteen combinations are also divided into two subcases ((c) and (d)). The subcases (c) and (d) further contain four and twelve combinations, respectively. The subcases are explained in detail in Table 5.

Subcase (c): When $x_{best}$ is not equal to $x_{worst}$ and $x_{worst}$ is equal to $x_{middle}$, then the obtained results are equal to $x_{best}$.

Subcase (d): If any case arises other than (c), then the obtained results are equal to $x_t$ by taking −2 and −1 as 0 and 2 and 3 as 1.

The mathematical formulation of Case 2 is given as

$$x_{tp}^{new} = \begin{cases} x_{best}; \text{if } x_{best} \neq x_{worst} = x_{middle} \\ x_t; \text{Otherwise} \end{cases}$$

The Pseudo-code of BGSK is presented in Fig. 5.

1. **Begin**
2. **Initialize the value of parameters** $Gen^{max}, NP, k_r, k, p$
3. **Initialize the generation** $(G = 0)$
4. **Create binary population using equation (23)**
5. **Evaluate** $f(x_t)$.
6. **For** $G = 1$ **to** $Gen^{max}$
7. **Compute the dimensions of both stages (Binary junior and senior gaining sharing stage)**
8. **Apply Binary Junior gaining sharing stage**
9. **Apply Binary Senior gaining sharing stage**
10. **Update the population**
11. **Select the global best solution**
12. **End for** $NP$
13. **End for** $G$
14. **End for Begin**

**Fig. 5** Pseudo-code for BGSK

# 7 Experimental Results

The TAP is solved using the proposed novel BGSK algorithm and the values of parameters are presented in Table 6. BGSK runs over personal computer Intel ® CoreTM i5-7200U CPU @ 2.50GHz and 4 GB RAM and coded on MATLAB R2015a. To get the optimal solutions, 25 independent runs are performed, and the statistical results are provided in Table 7, which includes the best, median, average, worst solutions and the standard deviations of BGSK. Moreover, Fig. 6 shows the convergence graph of the solutions of TAP using BGSK. From the Fig. 6, it can be observed that after the 18th iteration, it converges to the global optimal solution, which shows the robustness of the BGSK.

The current routine solution route provided by the health and safety agency management based on selecting the workplaces with the highest profit is to go from the headquarter to workplace number 3 with the highest profit over all other

**Table 6** Numerical values of parameters

| Parameters of BGSK | Considered values |
|---|---|
| NP | 800 |
| k | 10 |
| $k_r$ | 0.9 |
| p | 0.1 |
| $k_f$ | 1 |
| Max number of iterations | 50 |

**Table 7** Statistical results of TAP using BGSK

| Algorithm | Best (Maximum) | Median | Average | Worst (Minimum) | Standard Deviation |
|---|---|---|---|---|---|
| BGSK | 11.3 | 11.3 | 11.3 | 11.3 | 0.00 |

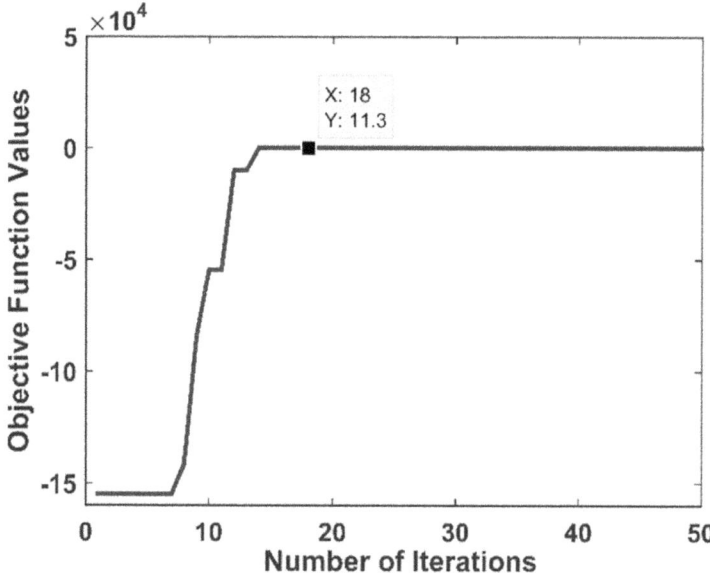

**Fig. 6** Convergence graph of BGSK

**Fig. 7** Optimum solution for the case study example

workplaces ($550), then to serve workplace number 2 with the next to the highest profitable one ($470). The total profit for this solution is $1020 with a total route length of 7.5 hours.

The obtained optimal solution from the proposed mathematical model is more profitable ($1130) by 10.8% higher, and a total route time length is 7.75 hours. This optimum solution is to start by visiting workplace 1, then workplace 2 and then workplace number 5 as indicated in Fig. 7.

# 8   Conclusions and Points for Future Research

The main conclusions for this paper can be summarized as follows:

1. A brand-new application problem known as the Traveling Advisor Problem (TAP) is introduced, in which an advisor must choose the applicant's workplaces he or she would visit in the most profitable manner. The occupational health and safety field's advisory expertise is where the issue primarily manifests itself.
2. The Traveling Advisor Problem (TDP) resembles the well-known Traveling Salesman Problem (TSP), but there are some key differences. These include the limited amount of route time, visiting only a portion of the locations, taking into account the time spent in customers' locations as a fundamental component, and the goal of maximizing total profit.
3. For the specified issue, a nonlinear binary mathematical model is developed. The suggested route's proposed positions are represented by the binary decision variables, and one of the constraints exhibits nonlinearity. The ultimate goal is to maximize the chosen route's overall profit.
4. A unique Binary GainingSharing Knowledge-based optimization technique (BGSK), which combines the two primary binary junior and senior gaining-sharing stages with the knowledge factor $k_f = 1$, is introduced to get the solution of the suggested TAP Nonlinear Binary Programming Model. The proposed technique is implemented with an augmented Lagrangian method to address the constraints in the BGSK.
5. A case study example is simulated and solved using the nonlinear binary mathematical model and the solution technique. The management's recommended solution, which was based on choosing the workplaces with the largest profit within the time limit limitation, is outperformed by the optimal solution, which is more profitable by 10.8%.
6. The outcomes obtained by BGSK demonstrate its robustness and convergence and demonstrate that it is capable of locating the overall TAP optimal solution. When compared to the management of the health and safety agency, BGSK produces better results.

The points for future researches can be stated in the following points:

1. To propose other mathematical models' formulations for the same problem starting with the design of the decision variables and compare the effectiveness of computations for each model.
2. To augment the proposed TAP with its variations as: TAP with Time Window (TAPTW), Stochastic Traveling Advisor Problem (STAP), TAP with multiple advisors ($m$TAP), Multi-objective Stochastic Multiple Traveling Advisor Problem (MOS$m$TAP) and other variations.
3. To apply the same problem formulation to other similar advisory fields that can show up in many other consulting domains like industry, agriculture, business, education, telecommunications, investing, quality assurance, social and

community services, pollution, medical, tourism, marketing, sales, advertising, sports, arts, cooking, and others.

4. To check the performance of the BGSK algorithm in solving different types of complex optimization problems, and further works can be investigated by the extension of BGSK.

## References

Agrawal, P., Ganesh, T., & Mohamed, A. W. (2021). A novel binary gaining–sharing knowledge-based optimization algorithm for feature selection. *Neural Computing and Applications, 33*(11), 5989–6008.

Agrawal, P., Ganesh, T., & Mohamed, A. W. (2022). Solving knapsack problems using a binary gaining sharing knowledge-based optimization algorithm. *Complex & Intelligent Systems, 8*(1), 43–63.

Anily, S., & Mosheiov, G. (1994). The traveling salesman problem with delivery and backhauls. *Operations Research Letters, 16*(1), 11–18.

Applegate, D. L., Bixby, R. E., Chvátal, V., & Cook, W. J. (2006). *The traveling salesman problem. Princeton Series in Applied Mathematics* (pp. 1–5). Princeton University Press.

Applegate, D. L., Bixby, R. E., Chvátal, V., Cook, W., Espinoza, D. G., Goycoolea, M., & Helsgaun, K. (2009). Certification of an optimal TSP tour through 85,900 cities. *Operations Research Letters, 37*(1), 11–15.

Aramgiatisiris, T. (2004). An exact decomposition algorithm for the traveling salesman problem with backhauls, Operations Research and Management Science Units, Department of Industrial Engineering, Kasetsart University, Bangkok, Thailand. *Journal of Research in Engineering and Technology, 1*(2), April–June 2004.

Awad, M. S. G., & Nour El-Din, A. A. (2018). *Egyptian law regulations in occupational safety and health*, https://doi.org/10.13140/RG.2.2.24128.38404.

Baldacci, R., Hadjiconstantinou, E., & Mingozzi, A. (2003). An exact algorithm for the traveling salesman problem with deliveries and collections. *Networks: An International Journal, 42*(1), 26–41.

Bektas, T. (2006). The multiple traveling salesman problem: An overview of formulations and solution procedures. *Omega, 34*(3), 209–219.

Bhusnure, O. G., Dongare, R. B., Gholve, S. B., & Giram, P. S. (2018). Chemical hazards and safety management in pharmaceutical industry. *Journal of Pharmacy Research, 12*(3), 357–369.

Boyle, T. (2015). *Health and safety: Risk management*. Routledge.

Carnell, M., & Nebosh, D. (2017). *Preventing harm in the workplace workbook*.

Claus, A. (1984). A new formulation for the travelling salesman problem. *SIAM Journal on Algebraic Discrete Methods, 5*(1), 21–25.

Coady, C., Feltham-Scott, D. A., Fewer, C., Osborne, C., & Taylor, K. (2015). *Building a safer tomorrow, workplace health, safety and compensation commission of newfoundland and labrador*, Workplace NL, San Jose, CA, USA.

Dantzig, G., Fulkerson, R., & Johnson, S. (1954). Solution of a large-scale traveling-salesman problem. *Journal of the Operations Research Society of America, 2*(4), 393–410.

Demiral, M. F. (2021). *Integer programming model for two-centered double traveling salesman problem*.

Droste, I. E. A. C. (2017). *Algorithms for the travelling salesman problem* (Bachelor's thesis).

Dumitrescu, I., Ropke, S., Cordeau, J. F., & Laporte, G. (2010). The traveling salesman problem with pickup and delivery: Polyhedral results and a branch-and-cut algorithm. *Mathematical Programming, 121*(2), 269–305.

El-Quliti, S. A., Ragab, A. H. M., Abdelaal, R., Mohamed, A. W., Mashat, A. S., Noaman, A. Y., & Altalhi, A. H. (2015). A nonlinear goal programming model for university admission capacity planning with modified differential evolution algorithm. *Mathematical Problems in Engineering, 2015*.

El-Qulity, S. A., & Mohamed, A. W. (2016a). A generalized national planning approach for admission capacity in higher education: A nonlinear integer goal programming model with a novel differential evolution algorithm. *Computational Intelligence and Neuroscience, 2016*.

El-Qulity, S. A. H., & Mohamed, A. W. (2016b). A large-scale nonlinear mixed-binary goal programming model to assess candidate locations for solar energy stations: An improved real-binary differential evolution algorithm with a case study. *Journal of Computational and Theoretical Nanoscience, 13*(11), 7909–7921.

El-Qulity, S. A., Mohamed, A. W., Bafail, A. O., & Abdelaal, R. (2016). A multistage procedure for optimal distribution of preparatory-year students to faculties and departments: A mixed integer nonlinear goal programming model with enhanced differential evolution algorithm. *Journal of Computational and Theoretical Nanoscience, 13*(11), 7847–7863.

Finke, G., Claus, A., & Gunn, E. (1983). A two-commodity network flow approach to the travelling salesman problem, Combinatorics, Graph theory and computing. In *Proc. 14th South Eastern Conf., Atlantic University, Florida*.

Fox, K. R., Gavish, B., & Graves, S. C. (1980). An n-constraint formulation of the (time-dependent) traveling salesman problem. *Operations Research, 28*(4), 1018–1021.

Gavish, B., & Graves, S. C. (1978). *The travelling salesman problem and related problems*.

Gendreau, M., Hertz, A., & Laporte, G. (1996). The traveling salesman problem with backhauls. *Computers & Operations Research, 23*(5), 501–508.

Gendreau, M., Laporte, G., & Vigo, D. (1999). Heuristics for the traveling salesman problem with pickup and delivery. *Computers & Operations Research, 26*(7), 699–714.

Gleixner, A. M. (2014). Introduction to constraint integer programming. In *5th Porto meeting on mathematics for industry*. Zuse Institute Berlin, MATHEON, Berlin Mathematical School.

Halse, K. (1992). *Modeling and solving complex vehicle routing problems* (Doctoral dissertation, Technical University of Denmark).

Kara, I., & Bektas, T. (2003, July). Integer linear programming formulation of the generalized vehicle routing problem. In *EURO/INFORMS joint international meeting, Istanbul*, July (pp. 06–10).

Manu, P., Emuze, F., Saurin, T. A., & Hadikusumo, B. H. (Eds.). (2019). *Construction health and safety in developing countries*. Routledge.

Marriott, M. C., & Schmidt-McCleave, R. (2018). *Safeguard Health and Safety Handbook 2019, omson Reuters New Zealand Limited*. Wellington, New Zealand.

Miller, C. E., Tucker, A. W., & Zemlin, R. A. (1960). Integer programming formulation of traveling salesman problems. *Journal of the ACM (JACM), 7*(4), 326–329.

Mohamed, A. W., Hadi, A. A., & Mohamed, A. K. (2020). Gaining-sharing knowledge based algorithm for solving optimization problems: A novel nature-inspired algorithm. *International Journal of Machine Learning and Cybernetics, 11*(7), 1501–1529.

Mohamed, A. K., & Mohamed, A. W. (2019). Real-parameter unconstrained optimization based on enhanced AGDE algorithm. In *Machine learning paradigms: Theory and application* (pp. 431–450). Springer, Cham.

Mohamed, A. W., Mohamed, A. K., Elfeky, E. Z., & Saleh, M. (2019a). Enhanced directed differential evolution algorithm for solving constrained engineering optimization problems. *International Journal of Applied Metaheuristic Computing (IJAMC), 10*(1), 1–28.

Mohamed, A. K., Mohamed, A. W., Elfeky, E. Z., & Saleh, M. (2019b). Solving constrained non-linear integer and mixed-integer global optimization problems using enhanced directed differential evolution algorithm. In *Machine learning paradigms: Theory and application* (pp. 327–349). Springer, Cham.

Mohamed, A. W., Sabry, H. Z., & Farhat, A. (2011, December). Advanced differential evolution algorithm for global numerical optimizatiom. In *2011 IEEE International Conference on Computer Applications and Industrial Electronics (ICCAIE)* (pp. 156–161). IEEE.

Mosheiov, G. (1994). The travelling salesman problem with pick-up and delivery. *European Journal of Operational Research, 79*(2), 299–310.

Oberlin, P., Rathinam, S., & Darbha, S. (2009, June). A transformation for a heterogeneous, multiple depot, multiple traveling salesman problem. In *2009 American control conference* (pp. 1292–1297). IEEE.

Onder, G., Kara, I., & Derya, T. (2017). New integer programming formulation for multiple traveling repairmen problem. *Transportation Research Procedia, 22*, 355–361.

Orman, A. J., & Williams, H. P. (2006). A survey of different integer programming formulations of the travelling salesman problem. *Optimisation, Econometric and Financial Analysis, 9*, 93–108.

Parker, L. (2017). *The early years health and safety handbook*. Routledge.

Pinter, C. C. (2014). *A book of set theory*. Courier Corporation.

Pop, P. C. (2007). New integer programming formulations of the generalized traveling salesman problem. *American Journal of Applied Sciences, 4*(11), 932–937.

Pureza, V., Morabito, R., & Luna, H. P. (2018). Modeling and solving the traveling salesman problem with priority prizes. *Pesquisa Operacional, 38*, 499–522.

Sacadura-Leite, E., Mendonça-Galaio, L., Shapovalova, O., Pereira, I., Rocha, R., & Sousa-Uva, A. (2018). Biological hazards for healthcare workers: Occupational exposure to vancomycin-resistant staphylococcus aureus as an example of a new challenge. *Portuguese Journal of Public Health, 36*(1), 26–31.

Salim, S. M., Romli, F. I., Besar, J., & Negin, O. A. (2017). A study on potential physical hazards at construction sites. *Journal of Mechanical Engineering (JMechE), 1*, 207–222.

Sarubbi, J. F. M., & Luna, H. P. L. (2007, April). The multicommodity traveling salesman problem. In *INOC–internacional network optimization conference, Belgian*.

Sawik, T. (2016). A note on the Miller-Tucker-Zemlin model for the asymmetric traveling salesman problem. *Bulletin of the Polish Academy of Sciences: Technical Sciences, 3*.

Sika Group. (2020). *Occupational safety and health, Sika Egypt website, 2020*. Found at https://egy.sika.com/content/egypt/main/en/group/Aboutus/sustainability/environment_and_safety/Safety.html

Silva, M. M., Subramanian, A., Vidal, T., & Ochi, L. S. (2012). A simple and effective metaheuristic for the minimum latency problem. *European Journal of Operational Research, 221*(3), 513–520.

Song, Y., Wu, D., Wagdy Mohamed, A., Zhou, X., Zhang, B., & Deng, W. (2021). Enhanced success history adaptive DE for parameter optimization of photovoltaic models. *Complexity*, 2021.

Vajda, S. (1961). *Mathematical programming*. Addison-Wesley.

Wong, R. T. (1980, October). Integer programming formulations of the traveling salesman problem. In *Proceedings of the IEEE international conference of circuits and computers* (Vol. 149, p. 152). IEEE Press Piscataway NJ.

# Data Envelopment Analysis for Healthcare Systems Assessment: Review and Applications from Tunisia

**Marwa Hasni, Safa Bhar Layeb ⓘ, Najla Omrane Aissaoui, and Aymen Mannai**

**Abstract** This chapter provides a critical review on the DEA method with a focus on its contribution to assess the performance of healthcare organizations around the world and specifically in Tunisia. To start with, we shed light upon the efficient border-based methods to help to systematically pinpoint the DEA among its counterparts. Afterwards, main key features of the DEA are discussed, and information on their theories, their corroborative evidence in the literature as well as directions for further enhancements are provided. Focusing on the healthcare field, this chapter also sketches the contributions of the DEA as an aid decision tool. In particular, we highlight the existence of two research strands, namely cross-sectional and overall based performance assessment of healthcare centers. Additionally, relevant gaps that need to be filled are stressed, most notably for studies dealing with the assessment of units within one healthcare center. Particular attention is also paid to the experiences of using DEA-based approach to assess cross-efficiency healthcare organization in Tunisia. Case studies as well as the lessons learned are reported.

M. Hasni
Higher Institute of Industrial Systems of Gabes, University of Gabes, Gabes, Tunisia
e-mail: marwa.gharbi@enit.rnu.tn

S. B. Layeb (✉) · A. Mannai
LR-OASIS, National Engineering School of Tunis, University of Tunis El Manar, Tunis, Tunisia
e-mail: safa.layeb@enit.utm.tn

N. O. Aissaoui
LR-OASIS, National Engineering School of Tunis, University of Tunis El Manar, Tunis, Tunisia

National Engineering School of Carthage, University of Carthage, Tunis, Tunisia

# 1   Introduction

The frontier-based approach is widely used for evaluating the efficiency of operational systems within various organizations (e.g., Cordero Ferrera et al., 2014). In short, efficient frontiers are curves that are obtained by plotting optimal points. In the relevant literature, a plethora of methods have been found to explore the frontier-based approach. In more details, these methods could be allied to either the parametric or the non-parametric approaches. Precisely, the parametric methods establish a comprehensive relation between the available inputs being at the entry of a system and the observed outputs, using a mathematical expression. The latter could be a production, a cost, or a profit function. Yet, the available parametric methods consist in the Stochastic Frontier Approach (SFA), the Distribution-Free approach (DFA), or the Thick Frontier Approach (TFA). Interested readers to each of these approaches are referred to the interesting works of Aigner et al. (1977), Berger (1993), and Berger and Humphrey (1991), respectively.

On the other hand, the non-parametric methods omit the use of explicit mathematical functions for modeling the relations between entrants and outputs. Substantially, these variants rely upon a comparative reasoning according to which the units of the system are evaluated given a particular behavior of their counterparts that are derived from the same sector. There are three main non-parametric methods, namely: the Data Envelopment Analysis (DEA) initially introduced by Charnes et al. (1978); the Free-Disposal Hull (FDH) developed according to the works of Thrall (1999) and Cherchye et al. (2000); and the Malmquist Productivity Index (MPI) presented by Worthington (1999), Grifell-Tatje and Lovell (1996).

Regarding the literature on healthcare applications, it could be stated that the DEA and the SFA are the mostly used methods. Technically, both methods consider the efficiency of the so-called Decision-Making Units (DMUs) as the quotient between its entrants and outputs. As such, the goal consists in evaluating health systems efficiencies given available input factors (e.g., availability of medical items, number of physicians) and exploring entrants for the purpose of optimizing observed outputs (e.g., the number of patients treated). It is worthy to mention that the DEA is more used than the SFA in the field of healthcare as it allows mainstreaming multiple inputs and outputs, and handles differences in weights and scales performances. Thereby, we carry out a systematic review that investigates the usage of the DEA for the measurement of healthcare systems efficiencies.

The rest of this chapter is structured as follows. Section 2 presents the DEA method and its main characteristics. Based on the new classification scheme, Sect. 3 reviews the main applications of the DEA within healthcare systems. Then, Sect. 4 is dedicated to report the case studies conducted in Tunisia. Challenges and opportunities are discussed in Sect. 5. Finally, Sect. 6 draws conclusions and avenues for future research.

## 2 Main Features of the DEA Approach

The Data Envelopment Analysis is a non-parametric quantitative technique that has been originally developed by Charnes et al. (1978). Their pioneer work introduced this method as a tool for measuring the cross-efficiency of the so-called decision-making units. Since, several works have extensively explored the DEA method to assess the effectiveness of various systems from many fields (e.g., Layeb et al., 2020a).

### 2.1 Main DEA Models

Given an initial set of homogeneous DMUs along with their underlying input and output factors, the DEA evaluates the relative efficiencies between these units by computing the ratio between the weighted sum of the outputs and weighted sum of inputs and reads as

$$
\text{Relative efficiency} = \frac{\text{Weighted sum of outputs}}{\text{Weighted sum of inputs}}. \tag{1}
$$

Considering $n$ DMUs, each having $m$ inputs and $s$ outputs, the relative efficiency score of DMU $p$ is obtained by solving the following model M1 (Charnes et al., 1978):

$$
\text{M1}: \max \frac{\sum_{k=1}^{s} \vartheta_k y_{kp}}{\sum_{j=1}^{m} u_j x_{jp}} \tag{2}
$$

Subject to

$$
\frac{\sum_{k=1}^{s} \vartheta_k y_{ki}}{\sum_{j=1}^{m} u_j x_{ji}} \leq 1, \forall i, \tag{3}
$$

$$
\vartheta_k, u_j \geq 0, \forall k, j, \tag{4}
$$

where $y_{ki}$ is the amount of the produced output $k$ per DMU $i$, $x_{ji}$ is the amount of the explored input $j$ per DMU $i$, $\vartheta_k$ is the weight of the input $k$, and $u_j$ is the weight of the input $j$. Typically, the fractional model M1 can be converted into a Linear program M2 that reads as

$$\text{M2}: \max \sum_{k=1}^{s} \vartheta_k y_{kp} \tag{5}$$

Subject to

$$\sum_{j=1}^{m} u_j x_{jp} = 1, \tag{6}$$

$$\sum_{k=1}^{s} \vartheta_k y_{ki} - \sum_{j=1}^{m} u_j x_{ji} = 1 \leq 0, \forall i, \tag{7}$$

$$\vartheta_k, u_j \geq 0 \forall k, j \tag{8}$$

Model M2 is run $n$ times to identify the relative efficiency scores of all DMUs. Each DMU selects input and output weights that maximize its score efficiency. In general, a DMU is considered effective only if it has a score of 1.

On the other hand, the application of the DEA method implies the specification of the following set of features:

- Identification of the type of scale returns
- Calculation of the efficiency score
- Identification of the reference peers (Benchmarking)

In the following, we provide detailed descriptions on each of these features.

## 2.2 Types of Scale Efficiency

There are three types of return to scale assumptions, namely constant, increasing and decreasing. Actually, Constant Return to Scale (CRS) means that efficiency is estimated, assuming that the outputs can increase by exactly the same proportion as the inputs. Whereas estimating efficiency in a Variable Return to Scale (VRS) case assumes that the increase in inputs does not result in a proportional change in outputs. Obviously, the performance variation of scales might be increasing, i.e., the increase in inputs results in a non-proportional but positive change of outputs, or decreasing, showing that the increase in inputs leads to a non-proportional but negative change of outputs.

## 2.3 Calculation of the Efficiency Scores

Linear programming is the underlying methodology that makes the DEA method particularly powerful compared to other efficiency measurement tools. In the existing literature, we distinguish four basic variants that ensure the calculation of the efficiency scores, as follows:

- Charnes, Cooper, and Rhodes's model (Charnes et al., 1978), usually called the CCR model, is the first model of the DEA method. It is usually represented as the fractional model M1 or the linear model M2. Besides, it commonly operates using the constant return on scale assumption.
- Banker, Charnes, and Cooper's model (Banker et al., 1984), usually called the BCC model, measures the efficiency using a convexity constraint that ensures that the composite unit has the same scale size as the evaluated unit. The DMUs with the lowest input levels or the highest output levels are assessed to be effective. The linear representation of the BCC model is quite similar to the CCR in model M2, while taking into account that the BCC variant operates under a VRS assumption.

At this stage, we shall point out that the CCR and BCC models can be either input-oriented or output-oriented. Precisely, an input-oriented model would provide information on the amendments on inputs that are required to allow the inefficient decision-making units to achieve efficiency. Conversely, an output-oriented model will provide the extent of increase of outputs to allow an inefficient unit reaches efficiency (Cooper et al., 2011).

- The additive model has been proposed by Charnes et al. (1989) by operating under a CRS assumption while combining both input and output orientations to reach an efficient point on the optimal frontier.
- The Slacks-Based Measure (SBM) based-model introduced by Tone (2001) enriches the additive counterpart by introducing a measurement that makes the efficiency evaluation senseless to the variations of the units of the inputs and outputs. Yet, this variant operates under a VRS assumption.

## 2.4 Identification of the Reference Peers (Benchmarking)

For each inefficient DMU, the DEA identifies a set of corresponding effective units that can be used as benchmarks for improvement. Although the DEA benchmarking makes it possible to identify targets for improvement, it presents some drawbacks. Indeed, an inefficient DMU may not be fundamentally similar to its benchmark regarding a number of key features. For example, considering an effective unit A against a non-effective unit B, whereby each of them has its own operational practices, then, unit A could not be a benchmark DMU for unit B.

# 3 Applications of the DEA in the Healthcare Field

During the last decades, the DEA has been widely used for the purpose of evaluating systems performances within several sectors, such as banking, healthcare, or education. Interested readers could find good reference in the work of Forker and Mendez (2001). The authors have conducted a system review and provided a classification on the available DEA-related studies within different fields.

The usage of the DEA for the purpose of evaluating healthcare systems performance has firstly been introduced by Nunamaker (1983) for measuring the efficiency of nurse units. This pioneer study has opened a new avenue for researches that consists in applying the DEA for evaluating the technical performances of hospitals in the USA and all around the world. In particular, the DEA has been used under three main headings:

- Relative efficiency evaluation between the DMUs
- Identification of the best performing DMUs
- Developing new ways for improving organizations' performances (i.e., productivity)

Given the extensive literature, a number of authors got interested in developing surveys with the aim to develop a comprehensive interpretation to the usage of the DEA in healthcare. Amongst, we cite the interesting review of Kohl et al. (2019) that cluster related works into four classes according to the following research questions: (i) overall efficiency measurement of DMUs, (ii) system analysis and development of new improving methodologies, (iii) resolving specific managerial issues, and (iv) post improvement analysis. Furthermore, Cantor and Poh (2018) have also proposed another review study that groups investigations according to the DEA variant being used. More recently, given the growing literature using hybrid DEA models, Hasni et al. (2021b) have introduced a systematic classification for the allied studies based on two main criteria, namely; the purpose of the research whether to test the overall performance of a healthcare system or not, and the impact of the input criteria on the observed performances.

It is worthy to note that all the existing review studies are either focusing on the study objective or on the technical features of the DEA variant being used. Henceforth, we point out a lack of a systematic review that combines both of these aspects. Such a work would be of interest to direct decision makers to select adequate DEA techniques given a particular research question. To fill this gap, we carried out an extensive review encompassing more than 50 key research studies. At the outcome, we develop and introduce a new multi-criteria-based classification illustrated by Fig. 1. To the best of our knowledge, this is the most general classification that has been developed to update the related literature so far and to help to accurately select research directions for further studies.

Going into more details, we first question the aggregation degree of the study to determine whether a macro or micro-level evaluation is conducted. A macro-level evaluation includes a set of healthcare systems, whereas a micro-level evaluation

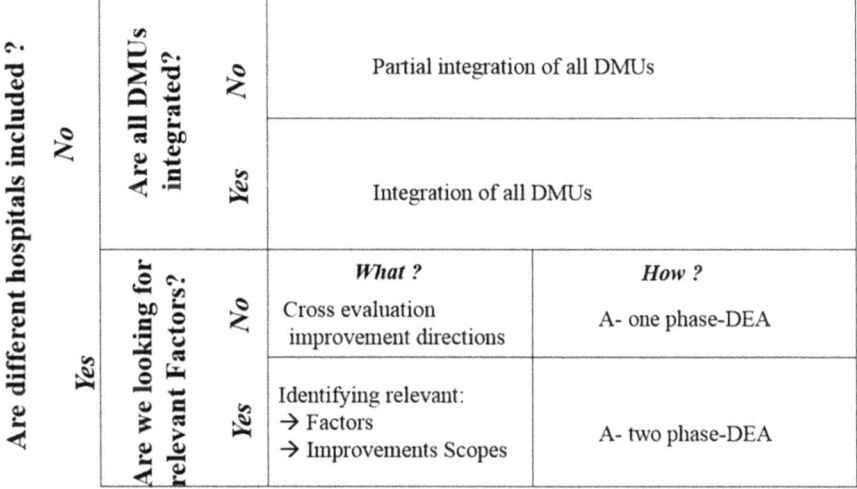

**Fig. 1** Classification of the DEA studies

deals with one healthcare system. In the context of micro-level evaluation, we distinguish two subsets of studies whereby the first aims at identifying relevant influencing criteria of an observed performance, and the second only establishes cross-evaluation to retrieve improvement scopes. Furthermore, we took forward this classification by tracking the explored DEA variant. Consequently, we noticed that the simple DEA model is dominant when dealing with research questions allied to the first subset a-two-phases DEA has been developed for the purpose of the studies belonging to the second subset. On the other hand, for the papers that deal with only one healthcare establishment, a sub-frontier could be highlighted to distinguish whether all DMUs are integrated or not.

In the following, we give detailed review on each group that has been identified in the proposed classification. Afterwards, we provide adequate synthesis in order to emphasize the main aspects underpinning the literature using DEA for healthcare assessment and shed light into gaps to be filled.

## 3.1 Macro-level-Based Studies

In this section, we investigate the literature on the available macro-level-based works evolving either a set of different hospitals or counterpart units that are selected from different establishments. As previously mentioned, a deep analysis to the underlying studies leaded us to distinguish two subsets whereby the first makes use of a one-step DEA model to measure efficiency in accurately allocating available resources; and the second, referred to as the two-phases DEA, investigates the impact of the factors on an observed performance.

### 3.1.1   One-Phase-DEA Models

This research strand relies on the DEA only for the measurement of the technical efficiency of healthcare systems and retrieves improvements directions. To start with, Grosskopf and Valdmanis (1987) used the CCR model with a CRS assumption to assess the technical efficiency of 82 hospitals, precisely 60 public and 22 private hospitals, in California. At the outcome, four entrants have been observed to significantly impact the observed performances, namely: the number of physicians, the working time of non-medical staff, the admission rate as well as the economic assets. The efficient units consist of the active healthcare, the surgical, the intensive care, the ambulatory, and the emergency care. Results have also shown that the public hospitals compared more favorable than private ones. Later, Wang et al. (1999) implemented the same DEA model to enact an investigation including four homogeneous metropolitan regions with at least two hospitals of active care units in 1989 and in 1993. The required data has been provided from the annual survey of the American Hospital Association (AHA). For that purpose, the inputs that have been used consist of the service complexity, the number of the available units offering diagnostic services and specific services, the number of the beds, the staff availability as well as the exploration cost. On the other hand, selected outputs include the hospital discharges and the number of consultations. The emanating results emphasized the poor performance of the expanded hospital markets. Indeed, the excessive number of the available workers resulted in an additional financial load of about 23 billions of dollars.

The CCR model with CRS assumptions has also been used by Chern and Wan (2000) to evaluate the impact of a newly implemented payment system on improving the operational hospitals' performance. For that purpose, the considered inputs are the number of the available beds, the total number of diagnostic and specific units, the number of non-medical workers, and the total expenditures, whereas the assessed outputs are restricted to the hospital discharges and to the number of consultants. The empirical results using data from 1984 to 1993 have outlined quiet similar performances. Henceforth, it was stated that the automation of the payment process is not of a direct relevance on healthcare performances.

Field and Emrouznejad (2003) analyzed the relative efficiency of 22 neonatal care units in Scotland using data records supplied from the national health services cost register that ranges from 1993 to 1994. To this end, both CCR and BCC models have been used. The selected inputs are the number of working days, the number of nurses, and the number of physicians. For the outputs, only the number of occupied beds has been considered. This study reveals a poor performance within the UK healthcare systems and points out potential productivity improvements of about 20%.

O'Neill and Dexter (2004) have applied an output-oriented CCR model to compute the technical efficiency of 53 non-metropolitan hospitals in Pennsylvania. The inputs are the numbers of beds, the surgical department for each hospital, the surgeons, as well as the hospital discharges. Their findings address issues in operating room management on the appropriate allocation of the additional resources to each surgical specialty. Then, Osei et al. (2005) have used the same model under both assumptions: constant and variable return to scale (i.e., CRS and VRS, respectively) to assess the performance of regional public hospitals performances and healthcare centers in Ghana. The models take as inputs: the number of doctors/ dentists, the number of the workers including the nurses, the number of early retired workers and the number of the available beds. At the outturn, evaluated outputs consist of the number of the maternal and child health visits, the births as well as the number of hospitalized patients. The numerical results have shown operational inefficiency in 47% of hospitals and 18% of healthcare centers. Henceforth, this study has put evidence on the versatility of the DEA in measuring inefficiency. Moreover, it provides the decision makers in Ghana with adequate directions to reform healthcare centers and hospitals along with a necessity to set up a monitoring system tracking the productivity growth, the efficiency in resources allocation, and the technical efficiency of all its health facilities.

To assess the technical and scale efficiencies of a sample of 37 peripheral health units in Pujehun district, Sierra Leone, Renner et al. (2005) have referred to an output-based BCC model. The inputs are the numbers of each of the medical and the non-medical staff, the amount of available resources and the total expenditures. Considering the outputs, light is shed onto antenatal and postnatal care: the number safely delivered new babies, the number of nutrition/growth follow-up visits, the number of family visits, the total number of people under 5 years old being vaccinated and the number of immunized pregnant women, and the total number of health education sessions conducted by home visits. The results have emphasized technical inefficiency in 22 units (59%) among the 37 tested ones with an overall efficiency score of 63%. Furthermore, 24 units, representing 65% of the considered sample, have been judged as scale inefficient with a mean score of scale efficiency of 72%. The public Namibian hospitals have also been benchmarked by Zere et al. (2006). In particular, the technical efficiency of 30 hospitals has been investigated using data records of the fiscal years from 1997 to 2001. For that purpose, three inputs and two outputs have been used, namely the total expenditures, the number of beds, the number of nurses as inputs and the number of ambulatory visits and the number of hospitalization days as outputs. The robustness of the DEA scores thus obtained has been evaluated using the Jackknife analysis. Thereby, a threefold aspects could best summarize the obtained results: (i) the overall technical efficiency score across all tested units of 75%, (ii) the technical efficiency frontier is drawn by less than 50% of the available hospitals, and (iii) the increasing return to scale are important driving factor to the observed inefficiency.

Hajialiafzali et al. (2007) have also conducted a variable return to scale (VRS) DEA that is input oriented to calculate the technical efficiency scores of Iranian hospitals that are affiliated to the Iranian social security organization, which is the second largest institutional of hospital care in Iran. The authors selected each of the numbers of physician, nurses, other staff workers and the beds as inputs. The outputs consist of the numbers of each of the surgical operations, overnight stays, treated patients as well as the medical procedures that are carried out to capture non-surgical procedures. The computational results suggest that among the 53 considered hospitals, 26 have been shown efficient. On the other hand, the remaining inefficient counterpart presented a 90% overall efficiency score. Henceforth, a reduction of the inputs with an amount of 10% could be a scope to reach efficiency. To end with, the authors have also provided a classification of the tested hospitals using an additional super-efficiency DEA model.

Nayar and Ozcan (2008) examined the technical efficiency and the service quality of 53 hospitals of Viriginia using the input-oriented DEA with a constant return to scale (CRS) assumption. A key aspect of this study is that it includes quality-based indicators to evaluate outputs. Going into more details, two investigations have been conducted separately. The first model considers four entrants and three outputs as follows: the inputs are the number of beds, the expenditures of the non-medical staff, the total asset, and the total number of worker, whereas the outputs encompass the total number of each of the medical consultations, the trainers and the adjusted total releases. The second model is similar to the first one but adds three quality-typed outputs, namely: the percentage of patients who received an initial antibiotic therapy, the percentage of patients assessed for oxygenation and the percentage of patients vaccinated against pneumococcal. At the outcome, the authors conclude that hospitals showing technical efficiency reveal satisfactory quality indicators. Interestingly, inefficient hospitals have also yield high-quality indicators. This triggers further avenues to search deeper insights on the influencing connection between technical and quality performances. A more extensive study using input-oriented DEA model with BCC assumption has been carried out by Mark et al. (2009) in the context of a research project on outcomes in nursing administration. The conducted study includes 226 healthcare units (i.e., care, surgical and medico-surgical) retrieved from 118 randomly selected hospitals. The considered inputs are the number of working hours of each of the nurses and the auxiliary nurses, the number of hours of work of support staff without permission, the number of beds and expenses. The Outputs are the adjusted landfills, the patient satisfaction, the number of medication errors and number of patient falls. This investigation has put evidence to the inefficiency of 60% of the considered units. In addition, case-based improvement directions have been provided. They consist in slightly reducing the working hours load and minimizing medication errors and falls. Patient security is another potential scope for enhancing the units' efficiencies.

Dash et al. (2010) applied an input-based DEA model with the BCC assumption to evaluate 29 district hospitals in Tamil Nadu State, India. The number of members of the workers and bed strength were used as inputs and the output data included are the number of outpatients, the number of hospitalized patients, the number of surgeries, the number of deliveries and the number of emergency cases. The emanating results show that 52% of the hospitals were technically effective as they had a 100% relative effectiveness score. Yet, the remaining 48% of hospitals were technically inefficient with an 82%-average efficiency score, leading to a potential improvement of 19% without being forced to reduce their current production levels. To do so, some effective hospitals could intentionally be used as reference. In the same line, Ketabi (2011) used the input-based BCC model to assess the effectiveness of 23 cardiac care units in Iran. The inputs considered in this study are the average number of active beds, the medical equipment, the number of the available staff (doctors, nurses and technicians) and technological capabilities. For the outputs, the author considered the percentage of bed occupancy, the average length of the period of stay, the survival percentage, and the performance ratio. Results have shown that Cardiac care units of 83.3% of the teaching hospitals and 60% of Ispahan's private hospitals are inefficient. This study also provides a classification of the effective units using the super-efficiency model of the DEA. An input-based BCC model has also been used by Gautam et al. (2013) to estimate the relative effectiveness of critical access hospitals in Missouri. The objective is to establish the trade between these variants of hospitals and the rural ones in the state. Indeed, rural hospitals are essential for access to health care and for their contribution to local economies. Many rural hospitals, especially the so-called critical access hospitals, must therefore increase their efficiency to ensure their viability. The outputs of this study are the number of stays, the number of ambulatory consultations, the number of non-emergency consultations, and the number of urgent consultations. The inputs are the number of employees, the number of beds and the load on workers. The results of this study indicate that rural hospitals compared more favorable than the CAH counterparts in terms of minimizing average service production cost.

Using data records from 2000 to 2010 and a dual DEA-based model, Arfa et al. (2017) have investigated the capacity utilization of Tunisian public hospitals. Recently, Ketabi et al. (2018) evaluated the performances of 24 emergency services in several hospitals in Isfahan, Iran. For that purpose, an input-oriented CCR model with a CRS assumption was implemented. The inputs consist in the numbers of each of the beds, physicians, nurses, and the medical equipment. On the other hand, the outputs are the number of outpatients, the survival rate, the average waiting time, and the satisfaction rate. This study highlighted the underperformance of 37% of Isfahan's emergency departments.

### 3.1.2   Two-Phases-DEA Models

Methods falling under this heading are described by two phases. In the first phase, either an input or an output-based DEA model is implemented under either CRS or VRS assumptions to compute efficiency scores of healthcare units. This phase could be assimilated to the studies reviewed in Sect. 3.1.1. However, a strand research study got interested in refining the obtained scores by grafting additional regression methods with a view to retrieve possible dependencies between the observed efficiencies and the patients' socio-demographic and clinical characteristics. This would be of interest to allow the inclusion of new influencing factors that have not been taken into account during the first phase of the analysis (Hollingsworth, 2016).

With reference to the relevant literature, we notice that both the bootstrapping and the regression approaches are the most common statistical tools that are referred to for the purpose of the second phase of the DEA. In particular, the bootstrapping approach is a resampling technique used to generate a big number of samples according to a particular experimental protocol (e.g., resampling with or without replacement) to help to generate an empirical sampled distribution of particular statistic of interest. In addition, sophisticated operations could be integrated into the resampling process (e.g., the jittering) to allow the appearance of new values within the sample. Regarding the regression method, as it is well known, it consists of a mathematical protocol aiming at determining the most influencing factor among the available explicative variables. It is the subset of variables to which the efficiency scores depend on the most. Moreover, these studies are motivated to answer the following research questions:

- What are the most and/or the least important factors that might describe observed efficiencies?
- What are the relative dependencies and interactions between the different factors to yield particular performance?
- What are the most available factors?

Various studies that could be allied to this research class are reviewed in the following. To start with, Grosskopf et al. (2004) evaluated the competitive effects on teaching hospitals operating in the USA. They analyzed the relative technical and scale efficiencies over a sample of 254 teaching hospitals. The DEA models being used in the first phase are each of the input-oriented CCR and BCC models with both CRS and VRS assumptions. The inputs are the numbers of each of the fully equipped beds, the doctors, the nurses, the medical residents and the number of other staff. The considered outputs are the outpatient visits to the hospital and emergency department, the number of patients, and the number of transactions. In the second phase, the authors used regression analysis in a bivariate context. They found that competition, measured by the number of care contracts managed per hospital and the number of patients covered by these contracts per hospital, has positive effects on teaching hospitals. Consequently, they proved that as far as competitiveness is triggered, the relative efficiency of the teaching hospital is likely to be enhanced.

Lee et al. (2008) investigated the relationship between the specialization index that measures the extent of specialization and the efficiency of Korean hospital services. Hospital specialization was measured using the theory index information based on empirical diagnoses. More precisely, the first phase used an input-oriented DEA under both assumptions CRS and VRS. The number of hospitalized patients and the number of visits have been considered as outputs, whereas the number of beds, the number of doctors and the number of nurses shaped the input set. In the second phase, a multiple regression model was developed to identify the most influencing factors on the specialization status of hospitals and their effectiveness. The main findings suggest that hospitals with different levels of specialization in services were more likely to be effective.

Kazley and Ozcan (2009) used a two-step DEA method to analyze the relationship between the use of electronic medical records and the change in efficiency over time in a national sample of care hospitals. The sources data of this study include the American hospital association and the health information management systems.

Firstly, the authors implemented an output-oriented DEA method with a CRS assumption. Inputs of the DEA model are the number of non-medical staff, the number of beds, the capital and the operating expenditures that are independent from labors. Meanwhile, outputs include the adjusted admissions for customer groups, and the outpatient visits. To determine the relationship between the use of numerical medical records and efficiency, logistic regression was then used. The authors found restricted evidence that numerical medical records could improve the efficiency of hospitals. Small hospitals can benefit from the use of electronic medical records, but medium-sized and large hospitals have been found to be unresponsive to variations of this factor.

Blank and Valdmanis (2010) described the effectiveness of 69 Dutch hospitals using a two-phase-based approach. In particular, their analysis focuses on the explanation of inefficiencies according to the environmental factors of each hospital. In the first stage, an input-oriented CCR model with a CRS assumption has been developed. The input set includes the numbers of each of the administrators, the nurses, the paramedics, the number of other staff and the number of equipment. Whereas the considered outputs are the landfills and the number of consultations for the first time. In turn, for the second stage, the bootstrap algorithm has been explored for the regression. At the outcome, the first stage DEA indicated that the profitability of general hospitals is between 62% and 100%. For its part, the bootstrapping approach has shown that the cost efficiency scores are affected by the operating environment. More specifically, the number of physicians significantly contributes to cost ineffectiveness.

The efficiencies and the perceived quality performances of rural and urban hospitals are compared in the work of Garcia-Lacalle and Martin (2010). More precisely, their study focuses on the Andalusia health service in Spain, which has implemented a policy of freedom and a reimbursement system based on hospital performance. The authors used a two-step DEA approach to compare 27 rural hospitals during years 2003 and 2006. In the first stage, they chose the input-oriented CCR method with a CRS assumption and inputs as the number of beds, the number of doctors, and the number of nurses. Instead, each of the numbers of stays, ambulatory patients, emergency visits, diagnoses, and operations for each hospital are retained for outputs. In the second stage, a multidimensional scaling analysis has been carried out to provide a graphical representation of the relative position of all hospitals with four quality indicators, namely the satisfaction of medical staff, the satisfaction of staff care provider, the satisfaction with the care provided and the satisfaction with the rooms and facilities. The results show that rural and urban hospitals achieve similar results in the efficiency dimension, whereas rural hospitals perform much better than urban hospitals in terms of patient satisfaction and when all dimensions are simultaneously considered, some rural hospitals have been found to performer better. In the same line, Bilsel and Davutyan (2014) have analyzed 202 Turkish rural general hospitals from 2006 onwards. The authors used both CCR and BCC models to measure technical effectiveness along with each of the scale efficiency and the regression analysis to identify critical areas for performance improvement. The inputs used in the first stage are the numbers of each of the specialists, the general practitioners, the nurses and midwives, the employees, the expenses and the beds. The outputs are the number of ambulatory patients, the number of hospitalized patients, the number of surgeries and the number of deaths divided by the total number of surgeries. Here, we shall point out that during the second phase, analysis was restricted to the inefficient units emanating from either the hospital or the rural district to help to better identify weaknesses. The emerging conclusions could be reported as scale inefficiency is concentrated in the smallest 96 hospitals, whereas a poor scale performance is observed in the largest hospitals. Furthermore, the technical inefficiency that has been observed in public hospitals could be linked to the behavior of some specialists who intent to attract patients into their private clinics.

Simões and Marques (2011) studied the impact of the congestion effect on Portuguese hospital services performance. To this end, the first stage of the analysis entails CCR models under CRS assumptions and BCC models under VRS assumptions to compute efficiency. Additionally, the authors selected the capital expenditures as well as other operating expenses as inputs. Alternatively, each of the numbers of patients, emergency department visits, hospital visits and outpatients have been designed as outputs. The emerging results show that 11 out of the 68 considered hospitals are effective in the CCR model, whereas the BCC model exhibits the outperformance of 26 among them. Indeed, the BCC model demonstrates more effectiveness than the CCR. More specifically, the overall efficiency outlined with the BCC model is of about 86%, whereas that yield by the CCR is of about 74%. Henceforth, it could be stated that, on average, Portuguese hospitals could reduce their inputs by 14% under the BCC model (or 26% under the CCR model). Here, we shall note that these recommendations are not controversy and applying one of them, according to the retained model, would produce the same performances (i.e., patients, emergency visits, outpatient visits). For the purpose of the second phase of the approach, the authors used a double-Bootstrap procedure to assess the impact of the operating environment on efficiency. An estimate of the congestion inefficiency was then obtained, to let conclude a significant inefficiency in more than 50% of the major hospitals.

The relative efficiency of hospitals in China from 2002 to 2008 was investigated by Hu et al. (2012) using an output-oriented BCC model with VRS assumptions. This study is motivated by an ambition to highlight the impact of health insurance reforms that have been applied by the New Rural Cooperative Medical System (NRCMS) on healthcare systems efficiency. In the first stage, each of the numbers of the ambulatory patients, the inpatient visits and the patient deaths have been selected as outputs. Conversely, the number of doctors, the number of technicians, the number of other non-medical staff, and the value of fixed assets shaped the input sets. The empirical findings put evidence on a moderate efficiency of hospitals varying at last by 4%, with a slight improvement trend of about 14% during the sampling period. To get deeper insights on the factors that might explain the observed efficiency, a regression analysis has been implemented in the second stage. The findings stressed the importance of the for-profit and high-tech hospitals' quality in improving technical efficiency and emphasized a negative relationship between Government subsidies and efficiency for coastal regions. Yet, the medical reform of NRCMS has been shown of a significant effect in improving efficiency, especially for regions not coastal. In the same line, Chaabouni and Abednnadher (2012) have also conducted a two-stage DEA method to evaluate the technical efficiency of public hospitals in Tunisia during the period 2000–2007. Nedelea and Fannin (2013) examined the impact of a medical program, referred to as the Critical Access Hospitals (CAH), on hospital efficiency using a two-step DEA method. The objective of the first phase is to estimate the cost, the technical and the locative efficiency scores of a sample of rural hospitals. Densities Efficiency scores of prospectively paid CAH and rural hospitals are estimated and compared using a non-parametric core density estimator and a test-based on the bootstrap. In the second phase, the efficiency scores are regressed on the variables using bootstrapped truncated regressions. The results showed the preference of the CAH in terms of cost efficiency over the rural hospitals. Interestingly, quiet similar technical the CAH before and after their conversion, a slight decrease in benefits has been recorded against a modest improvement in technical efficiency. Thereby, it could be concluded that the CAH program has been able to realize cost-effective efficiency compared to that yield by pre-paid rural hospitals. However, this contribution remains unimpressive since no significant increasing in technical efficiency has been found.

Recently, three main research studies used the two-stage DEA approach to address the effect of industrial competition on technical efficiency for healthcare systems. To start with, we first report the work of Özgen Narcı et al. (2015), whose empirical analysis uses Turkish data to investigate the relationship between competition and hospital efficiency. Two particular inputs for this study consist of two reform programs that have been undertaken as part of the so-called health transformation program in Turkey. Precisely, Özgen Narcı et al. (2015) had explored data records of 1103 public and private general hospitals opened until 2010 in Turkey. Technical effectiveness of hospitals was estimated using an input-oriented BCC model. Yet, five distinguished factors have been considered for inputs and outputs, respectively. Indeed, the number of outpatient visits, the number of emergency visits, the number of surgeries and number of hospital days were used for outputs, whereas the number of beds, the number of specialists, the number of general practitioners, the number of nurses, and the number of all other staff were chosen for inputs. In the second stage, the Tobit regression multivariate was used to study the relationship between competition and efficiency along with controlling the effects of supply and demand characteristics on market and hospital characteristics. The results showed that 17% of hospitals were technically effective, while the regression analysis highlighted that the competition extent between hospitals was not statistically correlated to the technical efficiency of hospitals. The second prominent research study was conducted by Nedelea and Fannin (2017) to evaluate the difference in profitability between two groups of hospitals in the USA, as their previous work (Nedelea & Fannin, 2013). For that purpose, a two-stage DEA approach has been implemented according to the following methodology. Initially, Nedelea and Fannin (2017) used an output-based BCC model taking as inputs each of the staff number, the total expenditures and the capital of the hospital. Outputs were the number of outpatient visits, admissions, stays, emergency department visits, outpatient surgeries, and deliveries. Afterwards, the authors carried out a comparative analysis evolving a density analysis with bootstrap truncated regression, Tobit and the stochastic boundary analysis. At the outcome of the first stage, the DEA-based results showed that, on average, the CAH were less cost-effective than non-CAH rural counterparts. These findings have also been stressed by the results yield by the analysis of the density of the DEA-based cost efficiency scores. The marginal effects of variables using the bootstrap truncated regression, Tobit and the stochastic frontier analysis, suggest that CAH are less cost-effective than rural hospitals, although Tobit revealed statistically insignificant. To end with, Kang et al. (2017) evaluated the technical and scale effectiveness of 976 US emergencies in 2013. They used a two input-oriented CCR and BCC models in the first stage. Whereas in the second stage, they used multivariate logistic regression to study significant exogenous factors affecting the technical efficiency of emergency services. The input set consists of the number of beds and the work hours of physicians and nurses. Likewise, the considered outputs are the number of visits and the number of patients who have not completed treatment. The results recommended reengineering the emergency processes to allow optimal usage of input resources. Further, the logistic regression analysis demonstrated that additional functional areas in the emergency department, the length of stay and the percentage of patients arriving by ambulance were closely associated with the technical effectiveness of the emergencies.

## 3.2   Conducted Studies Within a Single Hospital

### 3.2.1   Integration of All DMUs

This section reviews research studies that consider the services of the same hospital as Decision-Making Unit (DMU). It worthy to notice that Hofmarcher et al. (2002) were the first to investigate the effectiveness of hospital units using the DEA. More precisely, they compared the technical efficiency of 31 units of an Austrian hospital, using data records from 1994 to 1996. To do so, individual efficiency scores of evaluated units and their respective trend are tracked using an input-oriented BCC model with variable scale performance. The model takes as inputs the numbers of each of the medical workers, the paramedics and the administrative staff, and is meant to track each of the number of holidays and the obtained credit points as key indicators for efficiency. The results relying upon the number of taken holidays yield an average level of efficiency of 96%, while those given by considering the second tracking criterion, the obtained credit points, brings back efficiency to an average value of 70%. Later, Wang and Yu (2006) presented a DEA-based tool to identify relative efficiency performance across four departments of the Beijing Teaching Hospital. As such, an input-oriented CCR model with CRS assumption has been used. The inputs for this study are the number of beds and the amount of drug being used. On the other hand, outputs consist of the turnover, the numbers of outpatients, inpatients, surgical operations, scientific publications and learners. Lyroudi et al. (2006) focused on assessing productivity of 10 units in a public hospital of Thessaloniki in Greek from May 2002 until January 2003 with a one-month scale. The DEA model being used integrated Malmquist with the aim to analyze the units' productivity. The data were drawn from the monthly reports of each unit and encompass the number of beds, and different cost components namely the operating, the pharmaceutical, the medical and non-medical supply, as well as personnel costs. These are taken as inputs, whereas the outputs are given by the number of stays, examination revenues, transactions, the hospital expenses, and the laboratory revenues including test counts and other revenues. Numerical results indicated that despite observed improvements within units efficiency, this indicator is yet to be devised as it has revealed of a high variability across months that is stressed by the increase of the competitiveness.

In Saudi Arabia, Al-Shayea (2011) studied the performance and the effectiveness of nine departments at the King Khalid Teaching Hospital using an input-oriented DEA method with a CRS assumption. The considered departments are surgery, medicine, pediatrics, obstetrics and gynecology, orthopedics, primary care, accidents, psychology and emergencies, whereas the data consist in the monthly data records of the year 2010. The set of inputs retains the total salary for physicians and nurses, whereas outputs are drawn from the number of served patients, the productivity of beds and the average turnover. The results show that orthopedics, primary care and psychology departments maintained a 100% efficiency score during the year.

The work of Sebetci and Uysal (2017) analyzed the technical efficiency of the departments of the Adnan Menderes teaching hospital in Turkey. The data were obtained from statistical records related to the year 2014. The authors proposed three models with a view to assess respective impact of input and/or outputs on the resulting efficiencies. First, the BCC model with a VRS assumption has been used to minimize the following inputs: the expenditures, the deficits of packages and the deductions from social security institutions. In turn, income was defined as production output. This has shed light to the best performing department, namely the orthopedic department, and recommended it to be a benchmark for enhancing the remaining departments. Then, the CCR model with a CRS assumption has been designed where the teaching members, the research assistants, the number of rooms and the number of beds in the services were defined as inputs, and, the total number of patients, the total number of beds, and income, as outputs. Consequently, the departments of emergency and child and adolescent psychology outperformed the other departments. Lastly, the input set of the second counterpart is maintained, but outputs have been altered into the total number of patients assigned, the total number of surgical procedures and the revenues. This amendment has led to different performances since the department of thoracic surgery has emerged as the most effective department.

### 3.2.2  Partial Integration of DMUs

Now, let us turn our attention to the works proposing DEA analysis that consider groups of specialized units within the same healthcare system as DMUs. We begin by pointing out the work of Magnussen and Nyland (2008) that addressed how to apply input-oriented BCC and CCR models not only to assess relative departments' efficiency within a particular hospital, but also to propose a classification of them into more homogeneous groups. This would be of a direct practical relevance for decision makers to retrieve conclusions that are more comprehensive and draw adequate decisions. Cost-typed inputs have been fed to both models, namely the operating and the landfills and equipment costs. The considered outputs are the number of hospitalized patients, long-term care and stays. Subsequent results show significant boarders between five homogeneous groups of units, leading to conclusive performances. In addition, the authors assert the impact of the nature of the DEA variant being applied on the results. This goes in accordance with different similar studies in this research area. Precisely, a similar study has been conducted by Ancarani et al. (2009), who presented a model reflecting the relationships between the decision-making process within hospital units and their technical efficiencies. In order to test the proposed model, a two-stage investigation approach was adopted. Firstly, the technical efficiency of the rooms belonging to a large Italian hospital company was calculated using the DEA method. The dataset included 48 units out of a total of 81. It worthy to mention that the gynecology and obstetrics, critical care, emergency, and intensive care units were excluded from the analysis. In fact, gynecology and obstetrics unit were omitted due to the peculiarity of their admission and discharge records. Besides, intensive care, emergency care and pediatrics units were excluded because of the intensity of their resource utilization. Consequently, hospital departments have been divided into four main DMU groups: medical, surgical, inpatient and outpatient. Then, Ancarani et al. (2009) have applied CCR and BCC models with respective CRS and VRS assumptions. Data from available files are used to calculate the efficiency scores. For the inputs, the authors used the number of beds, doctors, non-medical staff, the changes in the use of operating rooms, i.e., each shift lasted six hours, and the maintenance costs of medical equipment. In the second stage, efficiency scores were lowered with the Tobit model on a set of variables capturing the internal objectives and managerial actions, as well as the reorganizations imposed by the central management of the hospital. The answers to a questionnaire administered to department heads were used to decide on the independent variables. Numerical results showed that exogenous reorganization processes and internal decisions of the units are important driving factor to the observed production efficiency.

In Alzahra hospital of Isfahan, Iran, Ketabi et al. (2015) have investigated empirical performance measurements for the management of the operating. Precisely, the authors compared the technical effectiveness of 11 attached specialties during the 2009. For that purpose, an output-oriented BCC model with VRS assumptions was applied to estimate cross-efficiency. In addition, the super-efficiency ranking model was used to rank the DMUs, along with pre-specified threshold scores for ineffective specializes of 0.96. The analysis also provided information on potential scopes for particular hospitals to enhance their respective surgeries schedules. In the same line, The study of Nakata et al. (2015) aimed at determining whether the current reimbursement system of surgeries in Japan reflects the use of resources following the revision of the 2014's fees schedule. The authors collected data on all Surgery at Teikyo Teaching Hospital from April, 1 to September, 30, 2014.They defined the DMUs as the hospital surgical units. The inputs were defined as the number of physicians who helped with the surgery as well and the time of the operations. The surgical revenues were the only considered output. The surgery departments concerned with this study were thoracic, neurosurgery, cardiovascular, obstetrics and gynecology, general surgery, orthopedics, urology, emergency, and plastic surgery. The efficiency scores for each surgical specialty were significantly different. This showed that the reimbursement scales for Japanese surgeries are yet to be devised in a way to reflect the use of resources. Remaining on assessing the performance surgical departments, another study were conducted by Girginer et al. (2015) within a public hospital in Eskisehir, Turkey. The authors combined the DEA to the Gray Relational Analysis (GRA). For the DEA, they used the BCC model with VRS assumptions. The Inputs are the total number of doctors, the number of other health workers, the bed turnover rate, and the bed occupancy rates. The considered outputs are the total number of outpatients, and the total number of operations. The DEA method showed that all services were significantly effective, with the exception of cardiovascular surgery and plastic departments. In fact, the GRA was used to classify the eight units with equal efficiency 1 and to determine the effects of the variables on performance of services.

## 4 Applications in Tunisia

Let us begin by mentioning three aforementioned studies on performance assessment of healthcare organizations conducted within African countries. Precisely, Osei et al. (2005) have used the model proposed by O'Neill and Dexter (2004) to assess the performance of regional public hospitals performances and healthcare centers in Ghana. Renner et al. (2005) have investigated the technical and scale efficiencies of 37 peripheral health units in Pujehun district, Sierra Leone. Furthermore, Zere et al. (2006) have assessed the technical efficiency of 30 public Namibian hospitals. In their recent work, Top et al. (2020) have investigated the technical efficiency of 36 African countries' healthcare systems. They found that about 60% of them are efficient and concluded that since African countries are economically depressed, their health systems are unfavorably affected.

Now, let us focus on the case studies conducted in Tunisia. Chaabouni and Abednnadher (2012) applied a two-stage DEA method to evaluate the technical efficiency of ten public hospitals in Tunisia during the period 2000–2007. They used a two-stage procedure whereby, the first stage implements an output-oriented BCC with VRS model to calculate efficiency scores. In the second stage, the authors focused on the identification on the most influencing factors to the observed performances. The experimental design encompasses historical data supplied by the Tunisian ministry of public health as well as the reports of the national institute of statistics of Tunisia. The set of inputs used for the purpose of the first stage includes the number of beds, physicians, nurses, dentists, pharmacists, and non-medical staff. For the outputs, the number of outpatient visits, admissions, and days after admission are considered. The efficiency scores yield by the first stage analysis are correlated with the explanatory variables used in the second stage as the logarithmic of the hospital expenses, the bed occupancy rate and the variable regional indicator. This analysis asserts no significant improvement to the healthcare efficiency during the years 2000–2007, leading to a fearful situation that is threatened to digress in performance.

Arfa et al. (2017) measured capacity utilization of Tunisian public district hospitals using data records in 2000 and in 2010. The authors occurred to the dual DEA method. Additionally, they implemented input and output restrictions to empower the empirical measurement of the capacity utilization. In particular, they added priorities in terms of input and output costs that are economically significant. The findings of this study suggest that the public sector hospitals do not operate in full capacity as the capacity utilization indicator has been observed to dramatically decrease from 101% in 2000 to 94% in 2010. More specifically, the public district hospitals in Tunis region have the highest capacity utilization rates, whereas the south-west regions have the lowest rates. Besides, they show potential amendments by 23% in 2010, if adequate improvements are applied. More recently, Abdelfattah (2021) have assessed the efficiency of 32 Tunisian regional hospitals using a neutrosophic DEA-based approach that could encounter missing and incomplete data.

In June 2019, the Tunisian ministry of health launched the annual breast cancer screening campaign, whose objective was the early detection of breast cancer in 500,000 women aged 35 years and older. In October 2019, it was reported that only 250,000 women actually performed the physical examination, of which 23,000 were invited to have a mammography, but that only 3400 women actually performed this examination. Because of these results, an efficiency assessment study was therefore conducted by Layeb et al. (2020b). A DEA-based approach was applied to evaluate the performance of the screening centers of the 24 governorates of Tunisia. The output of this work is to make recommendations based on best practices in resource allocation for improvement in order to ensure the success of the next breast cancer screening campaign.

Recently, several studies were conducted in Charles Nicolle university hospital. It is a sizable Tunisian public multi-specialty teaching hospital located at the Tunisian capital Tunis, where almost all medical specialties are covered. By its nature as a university hospital, 25 care units accommodate the learning process. Hasni et al. (2021a) have applied a new clustering-based DEA approach that clusters DMUs, herein the 25 care units, by evaluating the production function of the related features of the learning process. Experiments have shown that the Departments of pediatrics, gynecology B and the dentistry are effective. Despite that a plethora of studies dealt with teaching hospital units around the world (e.g., Grosskopf et al., 2004; Wang & Yu, 2006; Al-Shayea, 2011; Ketabi, 2011; Nakata et al., 2015) in evaluating healthcare units, they ruled out the evaluation of the teaching processes. To the best of our knowledge, a first attempt addressing this research question is carried out in Hasni et al. (2021a) within Charles Nicolle hospital. For the ambulatory care departments, Omrane et al. (2019) have investigated generating performance empirical measures using recorded data for the period 2014–2016. More precisely, the authors applied a BCC input-focused model that considers VRS assumptions. Numerical results revealed that only four units were found to be relatively effective over the three years. It is worth mentioning that the output of this study is considered helpful for the hospital decision makers to determine by how much outpatient units could decrease human resources while fulfilling their tasks to avoid wastage of those limited resources. To end with, Hasni et al. (2021b) got inspired by the work of Girginer et al. (2015) to develop and introduce a hybrid DEA-GRA model that is dedicated to a macro-level evaluation of the 63 care units Charles Nicolle hospital. The idea is to identify similar processes within units as well as their respective potential amendments using the DEA. Then, these findings are taken forward to be refined using the GRA. The authors start by splitting the care units into similarly four operating processes, namely: the emergency, the surgical, the hospitalization and the ambulatory processes. Then, a DEA-based investigation was carried out to identify efficient units within each process and to provide accommodating tractable measures for the input variables. Then, the efficient units are scanned by the GRA to highlight their preference order and thus retrieve the benchmarked units as well as the most influencing variables to prioritize. Computational results revealed that the overall efficiency scores of the care units fluctuated in the 2014–2016 period and did not follow a particular trend. In addition, the performance of the departments from the ambulatory and the hospitalization processes are less variable than those of the surgical and the emergency processes. Surprisingly, the number of physicians and nurses could be lowered to allow a productivity improvement for several departments. This pointed out a need to resize the resource requirements by means of interchangeability between units. This highlights the challenges related to the medical staff recruitment in Charles Nicolle hospital and other similar healthcare organizations in Tunisia.

## 5 Challenges and Opportunities

This chapter provides a review on the DEA and its role in enhancing healthcare centers efficiency that could represent a good reference for interested researchers to point out worthwhile research questions to investigate. Throughout the reviewed works, we notice that the first class representing studies encompassing different hospitals has been extensively studied in the literature by various authors, whereas fewer researches got interested to the remaining classes that restrict the focus on single hospital with either complete or partial integration of its units. Thereby, since the number of frontier-based investigations considering hospital units as DMUs are limited, it could be stated that poor information is available on single DMUs efficiency. This is yet to be investigated in order to allow the development of conclusive properties of each category. The observed shortage might further be explained by the fact that, given the shared nature of various input and output factors among most DMUs, most of authors prefer treating their respective total amounts rather than unit-based values. However, we shall point out that ignoring the difference of particular attributes (i.e., inputs/outputs) across DMUs can yield confusing conclusions between overall and per unit efficiencies.

It goes without saying that the DEA is a versatile approach that has been applied in a number of fields. In particular, we stress the common usage of each of the CCR and the BCC variants. Cantor and Poh (2018) pointed out that 91% of the available studies employing the DEA in the healthcare field refer to at least one these two models. Here we shall note that these models are not particular for healthcare and have been widely used in other efficiency evaluation contexts. Henceforth, an interesting research study would develop a DEA variant that is adapted to the healthcare fields. As such, we recommend reengineering adequate decision variables that should be closely tightened to the healthcare assumptions and characteristics.

A major drawback underpinning the DEA is that it does not distinguish efficient DMUs. Thereby, it would be confusing for decision makers to retrieve adequate directions for improvement if multiple and heterogeneous units reveal all efficient. In addition, a big number of efficient DMUs would be prohibitive to adopt and apply all their respective good practices as a scope for enhancing the reaming inefficient counterparts. Therefore, one would wish to further take forward the DEA analysis in order to emphasize differences among efficient units and perhaps classify them. According to this work, poor literature is available to deal with this research question. In addition, all available studies adopt a quiet similar reasoning that consist in grafting a ranking method to the DEA such as the super-efficiency (e.g., Ketabi et al., 2015; Ketabi, 2011; Hajialiafzali et al., 2007) or the GRA (e.g., Girginer et al., 2015; Hasni et al., 2021a, 2021b). Alternatively, other approaches such as decision tree-based methods could intentionally be investigated. These are powerful machine learning tools that ensure accurate classification with least data assumptions.

In the same vein, it is noteworthy to highlight the development of benchmark frontier that would serve as a reference for all healthcare-based studies. This would be of interest to ensure the applicability of the decisions that could be taken in the real world. Nevertheless, such an academic contribution requires an extensive and systematic empirical study that questions the availability and the pertinence of data records, especially in African countries.

# 6 Conclusion

This chapter reviewed the literature on the usage of the Data Envelopment Analysis with a focus to the healthcare field and its applications in Tunisia. In particular, we propose a new classification of the allied studies for a methodical understanding to the studies dealing with healthcare efficiency measurement using this approach. In short, the proposed classification grid splits the available studies into three major groups following their analysis aggregation degree whether a single or multiple healthcare systems are considered. The case studies as the lessons learned from applications in Tunisia were reviewed. Challenges of using DEA-based approach to assess cross-efficiency healthcare organizations as well research avenues were reported. For African countries, data availability and pertinence are still one of the most challenges to conduct accurate studies on performance assessment.

# References

Abdelfattah, W. (2021). Neutrosophic data envelopment analysis: An application to regional hospitals in Tunisia. *Neutrosophic Sets and Systems, 41*, 89–105.

Aigner, D., Lovell, C. K., & Schmidt, P. (1977). Formulation and estimation of stochastic frontier production function models. *Journal of Econometrics, 6*(1), 21–37.

Al-Shayea, A. M. (2011). Measuring hospital's units efficiency: A data envelopment analysis approach. *International Journal of Engineering & Technology, 11*(6), 7–19.

Ancarani, A., Di Mauro, C., & Giammanco, M. D. (2009). The impact of managerial and organizational aspects on hospital wards' efficiency: Evidence from a case study. *European Journal of Operational Research, 194*(1), 280–293.

Arfa, C., Leleu, H., Goaïed, M., & Van Mosseveld, C. (2017). Measuring the capacity utilization of public district hospitals in Tunisia: Using dual data envelopment analysis approach. *International Journal of Health Policy and Management, 6*(1), 9–18.

Banker, R. D., Charnes, A., & Cooper, W. W. (1984). Some models for estimating technical and scale inefficiencies in data envelopment analysis. *Management Science, 30*(9), 1078–1092.

Berger, A. N. (1993). "Distribution-free" estimates of efficiency in the US banking industry and tests of the standard distributional assumptions. *Journal of Productivity Analysis, 4*(3), 261–292.

Berger, A. N., & Humphrey, D. B. (1991). The dominance of inefficiencies over scale and product mix economies in banking. *Journal of Monetary Economics, 28*(1), 117–148.

Bilsel, M., & Davutyan, N. (2014). Hospital efficiency with risk adjusted mortality as undesirable output: The Turkish case. *Annals of Operations Research, 221*(1), 73–88.

Blank, J. L., & Valdmanis, V. G. (2010). Environmental factors and productivity on Dutch hospitals: A semi-parametric approach. *Health Care Management Science, 13*(1), 27–34.

Cantor, V. J. M., & Poh, K. L. (2018). Integrated analysis of healthcare efficiency: A systematic review. *Journal of Medical Systems, 42*(1), 1–23.

Chaabouni, S., & Abednnadher, C. (2012). Efficiency of public hospitals in Tunisia: A DEA with bootstrap application. *International Journal of Behavioural and Healthcare Research, 3*(3–4), 198–211.

Charnes, A., Cooper, W. W., & Rhodes, E. (1978). Measuring the efficiency of decision making units. *European Journal of Operational Research, 2*(6), 429–444.

Charnes, A., Cooper, W. W., Wei, Q. L., & Huang, Z. M. (1989). Cone ratio data envelopment analysis and multi-objective programming. *International Journal of Systems Science, 20*(7), 1099–1118.

Cherchye, L., Kuosmanen, T., & Post, T. (2000). What is the economic meaning of FDH? A reply to Thrall. *Journal of Productivity Analysis*, 263–267.

Chern, J. Y., & Wan, T. T. (2000). The impact of the prospective payment system on the technical efficiency of hospitals. *Journal of Medical Systems, 24*(3), 159–172.

Cooper, W. W., Seiford, L. M., & Zhu, J. (2011). Data envelopment analysis: History, models, and interpretations. In *Handbook on data envelopment analysis* (pp. 1–39). Springer, Boston, MA.

Cordero Ferrera, J. M., Cebada, E. C., & Murillo Zamorano, L. R. (2014). The effect of quality and socio-demographic variables on efficiency measures in primary health care. *The European Journal of Health Economics, 15*(3), 289–302.

Dash, U., Vaishnavi, S. D., & Muraleedharan, V. R. (2010). Technical efficiency and scale efficiency of district hospitals: A case study. *Journal of Health Management, 12*(3), 231–248.

Field, K., & Emrouznejad, A. (2003). Measuring the performance of neonatal care units in Scotland. *Journal of Medical Systems, 27*(4), 315–324.

Forker, L. B., & Mendez, D. (2001). An analytical method for benchmarking best peer suppliers. *International Journal of Operations & Production Management, 21*(1/2), 195–209.

Garcia-Lacalle, J., & Martin, E. (2010). Rural vs urban hospital performance in a 'competitive' public health service. *Social Science & Medicine, 71*(6), 1131–1140.

Gautam, S., Hicks, L., Johnson, T., & Mishra, B. (2013). Measuring the performance of critical access hospitals in Missouri using data envelopment analysis. *The Journal of Rural Health, 29*(2), 150–158.

Girginer, N., Köse, T., & Uçkun, N. (2015). Efficiency analysis of surgical services by combined use of data envelopment analysis and gray relational analysis. *Journal of Medical Systems, 39*(5), 1–9.

Grifell-Tatje, E., & Lovell, C. K. (1996). Deregulation and productivity decline: The case of Spanish savings banks. *European Economic Review, 40*(6), 1281–1303.

Grosskopf, S., Margaritis, D., & Valdmanis, V. (2004). Competitive effects on teaching hospitals. *European Journal of Operational Research, 154*(2), 515–525.

Grosskopf, S., & Valdmanis, V. (1987). Measuring hospital performance: A non-parametric approach. *Journal of Health Economics, 6*(2), 89–107.

Hajialiafzali, H., Moss, J. R., & Mahmood, M. A. (2007). Efficiency measurement for hospitals owned by the Iranian social security organisation. *Journal of Medical Systems, 31*(3), 166–172.

Hasni, M., Aissaoui, N., Layeb, S. B., & Manai, A. (2021b, May). A clustering-based data envelopment analysis for learning processes performance assessment in teaching hospitals. In *2021 1st International Conference On Cyber Management and Engineering (CyMaEn)* (pp. 1–6). IEEE.

Hasni, M., Layeb, S. B., Aissaoui, N. O., & Mannai, A. (2021a). Hybrid model for a cross-department efficiency evaluation in healthcare systems. *Managerial and Decision Economics, 43*, 1311–1329.

Hofmarcher, M. M., Paterson, I., & Riedel, M. (2002). Measuring hospital efficiency in Austria – A DEA approach. *Health Care Management Science, 5*(1), 7–14.

Hollingsworth, B. (2016). Health system efficiency: Measurement and policy. In Health system efficiency: How to make measurement matter for policy and management [Internet]. *European Observatory on Health Systems and Policies.*

Hu, H. H., Qi, Q., & Yang, C. H. (2012). Analysis of hospital technical efficiency in China: Effect of health insurance reform. *China Economic Review, 23*(4), 865–877.

Kang, H., Nembhard, H., DeFlitch, C., & Pasupathy, K. (2017). Assessment of emergency department efficiency using data envelopment analysis. *IISE Transactions on Healthcare Systems Engineering, 7*(4), 236–246.

Kazley, A. S., & Ozcan, Y. A. (2009). Electronic medical record use and efficiency: A DEA and windows analysis of hospitals. *Socio-Economic Planning Sciences, 43*(3), 209–216.

Ketabi, S. (2011). Efficiency measurement of cardiac care units of Isfahan hospitals in Iran. *Journal of Medical Systems, 35*(2), 143–150.

Ketabi, S., Ganji, H., Shahin, S., Mahnam, M., Soltanolkottabi, M., & Moghadam, S. A. H. Z. (2015). Surgical services efficiency by data envelopment analysis. *Benchmarking: An International Journal, 22*, 978–993.

Ketabi, S., Teymouri, E., & Ketabi, M. (2018). Efficiency measurement of emergency departments in Isfahan, Iran. *International Journal of Process Management and Benchmarking, 8*(2), 142–155.

Kohl, S., Schoenfelder, J., Fügener, A., & Brunner, J. O. (2019). The use of Data Envelopment Analysis (DEA) in healthcare with a focus on hospitals. *Health Care Management Science, 22*(2), 245–286.

Layeb, S. B., Aissaoui, N. O., Fatma, N. B., Frikha, M., Jemai, Z., Bohli, N., & Hamouda, C. (2020b, November). Efficiency analysis to evaluate a breast cancer screening campaign: A case study from Tunisia. In *2020 International Conference on Decision Aid Sciences and Application (DASA)* (pp. 412–414). IEEE.

Layeb, S. B., Omrane, N. A., Siala, J. C., & Chaabani, D. (2020a, December). Toward a PCA-DEA based Decision Support System: A case study of a third-party logistics provider from Tunisia. In *2020 4th International Conference on Advanced Systems and Emergent Technologies (IC_ASET)* (pp. 294–299). IEEE.

Lee, K. S., Chun, K. H., & Lee, J. S. (2008). Reforming the hospital service structure to improve efficiency: Urban hospital specialization. *Health Policy, 87*(1), 41–49.

Lyroudi, K., Glaveli, N., Koulakiotis, A., & Angelidis, D. (2006). The productive performance of public hospital clinics in Greece: A case study. *Health Services Management Research, 19*(2), 67–72.

Magnussen, J., & Nyland, K. (2008). Measuring efficiency in clinical departments. *Health Policy, 87*(1), 1–7.

Mark, B. A., Jones, C. B., Lindley, L., & Ozcan, Y. A. (2009). An examination of technical efficiency, quality, and patient safety in acute care nursing units. *Policy, Politics, & Nursing Practice, 10*(3), 180–186.

Nakata, Y., Yoshimura, T., Watanabe, Y., Otake, H., Oiso, G., & Sawa, T. (2015). Resource utilization in surgery after the revision of surgical fee schedule in Japan. *International Journal of Health Care Quality Assurance, 28*, 635–643.

Nayar, P., & Ozcan, Y. A. (2008). Data envelopment analysis comparison of hospital efficiency and quality. *Journal of Medical Systems, 32*(3), 193–199.

Nedelea, I. C., & Fannin, J. M. (2013). Impact of conversion to Critical Access Hospital status on hospital efficiency. *Socio-Economic Planning Sciences, 47*(3), 258–269.

Nedelea, I. C., & Fannin, J. M. (2017). Testing for cost efficiency differences between two groups of rural hospitals. *International Journal of Healthcare Management, 10*(1), 57–65.

Nunamaker, T. R. (1983). Measuring routine nursing service efficiency: A comparison of cost per patient day and data envelopment analysis models. *Health Services Research, 18*(2 Pt 1), 183.

O'Neill, L., & Dexter, F. (2004). Market capture of inpatient perioperative services using DEA. *Health Care Management Science, 7*(4), 263–273.

Omrane, N. A., Manai, A., Layeb, S. B., Siala, J. C., & Hamouda, C. (2019, March). Data Envelopment Analysis for Care Production Efficiency Measurement in Ambulatory Care Departments. In *2019 International Conference on Advanced Systems and Emergent Technologies (IC_ASET)* (pp. 102–107). IEEE.

Osei, D., d'Almeida, S., George, M. O., Kirigia, J. M., Mensah, A. O., & Kainyu, L. H. (2005). Technical efficiency of public district hospitals and health centres in Ghana: A pilot study. *Cost Effectiveness and Resource Allocation, 3*(1), 1–13.

Özgen Narcı, H., Ozcan, Y. A., Şahin, İ., Tarcan, M., & Narcı, M. (2015). An examination of competition and efficiency for hospital industry in Turkey. *Health Care Management Science, 18*(4), 407–418.

Renner, A., Kirigia, J. M., Zere, E. A., Barry, S. P., Kirigia, D. G., Kamara, C., & Muthuri, L. H. (2005). Technical efficiency of peripheral health units in Pujehun district of Sierra Leone: A DEA application. *BMC Health Services Research, 5*(1), 1–11.

Sebetci, Ö., & Uysal, İ. (2017). The efficiency of clinical departments in medical faculty hospitals: A case study based on data envelopment analysis. *International Journal of Computer Sciences and Engineering, 5*(7), 1–8.

Simões, P., & Marques, R. C. (2011). Performance and congestion analysis of the Portuguese hospital services. *Central European Journal of Operations Research, 19*(1), 39–63.

Thrall, R. M. (1999). What is the economic meaning of FDH? *Journal of Productivity Analysis, 11*(3), 243–250.

Tone, K. (2001). A slacks-based measure of efficiency in data envelopment analysis. *European Journal of Operational Research, 130*(3), 498–509.

Top, M., Konca, M., & Sapaz, B. (2020). Technical efficiency of healthcare systems in African countries: An application based on data envelopment analysis. *Health Policy and Technology, 9*(1), 62–68.

Wang, B. B., Ozcan, Y. A., Wan, T. T., & Harrison, J. (1999). Trends in hospital efficiency among metropolitan markets. *Journal of Medical Systems, 23*(2), 83–97.

Wang, R., & Yu, M. (2006, June). Evaluating the Efficiency of Hospital's Departments using DEA. In *2006 IEEE International Conference on Service Operations and Logistics, and Informatics* (pp. 1167–1170). IEEE.

Worthington, A. C. (1999). Malmquist indices of productivity change in Australian financial services. *Journal of International Financial Markets, Institutions and Money, 9*(3), 303–320.

Zere, E., Mbeeli, T., Shangula, K., Mandlhate, C., Mutirua, K., Tjivambi, B., & Kapenambili, W. (2006). Technical efficiency of district hospitals: Evidence from Namibia using data envelopment analysis. *Cost Effectiveness and Resource Allocation, 4*(1), 1–9.

# Artificial Intelligence and Operations Research in a Middle Ground to Support Decision-Making in Healthcare Systems in Africa

**Safa Elkefi ⓘD and Safa Bhar Layeb**

**Abstract** Advanced technologies have already taken the healthcare industry by storm. Artificial Intelligence (AI) is significantly becoming a trend that assists healthcare practices in many aspects of their patient care delivery by providing rich and flexible modeling perspectives of problems from the real world. It also allows the human expertise to be embedded in the solutions by supporting efficient reasoning mechanisms that are based on constraints reasoning and mixed-initiative frameworks. This chapter provides a critical review of AI use to solve combinatorial optimization problems of decision-making in healthcare. We focus specifically on the real-world benefits of these solutions and highlight their contribution to healthcare development by improving patients' experiences, doctors' experiences, and the work system environment. We cover the applications of AI in the world and then in Africa, focusing on the challenges and opportunities of implementation. For each of these aspects, information on their theories corroborated evidence in the literature as well as directions for further improvements will be provided.

## 1 Introduction

Operations Research (OR) may be defined as the analysis method that aids management makes decisions (Dean, 1958). It involves several scientific methods, techniques, and tools to solve systems' problems taking into consideration the systems' operations to provide solutions with an optimum control (Morse et al., 2003). As decision-making could be defined as the process of choosing between solutions to a problem respecting some metrics, OR concepts can then help systematically improve

S. Elkefi (✉)
Stevens Institute of Technology, Hoboken, USA
e-mail: selkefi@stevens.edu

S. B. Layeb
LR-OASIS, National Engineering School of Tunis, University of Tunis El Manar, Tunis, Tunisia
e-mail: safa.layeb@enit.utm.tn

© The Author(s), under exclusive license to Springer Nature Switzerland AG 2022    51
H. Masri (ed.), *Africa Case Studies in Operations Research*, Contributions to Management Science, https://doi.org/10.1007/978-3-031-17008-9_3

decision-making through efficient modeling techniques while accounting for relevant constraints providing management with quantitative information based on scientific methods of analysis (Capan et al., 2017).

OR is therefore used to coordinate activities, operations, and processes in organizations such as businesses and healthcare facilities. It, for example, helps ensure maximum health coverage for a given population or decide the number of beds, ambulances, or other resources needed based on many factors (Rais & Viana, 2011). Despite the considerable attention is given to strategic optimization problems in healthcare in the last decades, OR is now shifting more towards day-to-day management problems. The current focus of research could lead to many new avenues in how to solve the optimization issues related to planning, scheduling, and organizing preventive and diagnostic care (Royston, 2016).

On another side, Artificial Intelligence (AI) is currently among the major trending research topics and practice areas (Dornemann et al., 2020) that can mimic human cognitive functions (Jiang et al., 2017). It has witnessed massive progress with deep learning, computer vision, and robotics development (Reddy et al., 2019; Yu et al., 2018). It has also been proven to impact health systems through patient administration, clinical decision support, patient monitoring, and healthcare interventions, which play a central role in system development (Reddy et al., 2019).

In addition, AI takes advantage of a variety of knowledge representation forms to deal with the various problems from the real world. There are many examples of constraint-based models' representations, languages of programming for declaration and functioning such as Prolog and Lisp, Bayesian models, rule-based formalities, and other formalisms that cannot handle the sizes of problems realistically (Inozemtsev et al., 2017). Furthermore, the focus of OR has been more on traceable representations like linear programming formulations (Gass, 2003). While the expressive power of OR's rigid solving models is limited, AI's real-world representations techniques are more flexible (Kusiak, 1987). Actually, while AI usually emphasizes the qualitative aspects of the problems, OR stresses the quantitative ones. While AI solution methods focus on logic and inference, OR tools tend to be algorithmic. While AI focuses on managing reasoning knowledge, OR focuses on managing a procedural one (Holsapple et al., 1994). Despite these differences, there are similarities between the two disciplines. Most OR and AI specialists aim to solve complex combinatorial optimization problems, particularly for decision-making for scheduling, planning, and other logistical and organizational problems. They both involve the use of descriptive knowledge (data-based), including measurements, observations, or state-related beliefs (Gomes, 2000).

As solving optimization problems is a great challenge for human intelligence that involves the recognition and solution of one or more sub-problems, a major contribution of AI is in solving these problems with computer-based approaches that take into consideration account the data of the decisions made (Holsapple et al., 1994; Gomes, 2000; Kobbacy et al., 2007). Thus, a common ground of overlap and complementation between these two disciplines is expected. This overlap is referred

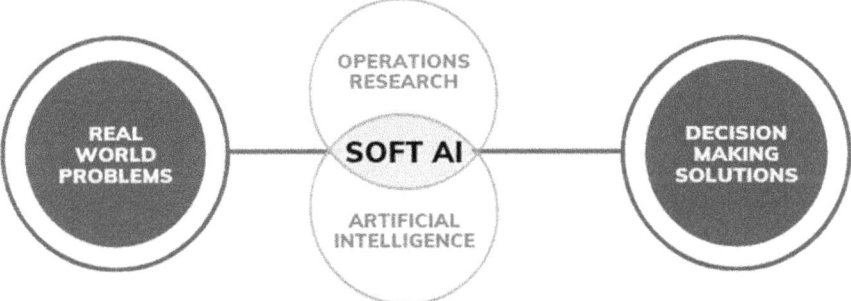

**Fig. 1** OR vs. AI for decision-making support

to as "*soft AI*," which uses many of the same search algorithms as OR, as shown in Fig. 1.

Over the past half-century, the two disciplines of OR and AI have evolved along parallel lines, each independent from the other (Holsapple et al., 1994). However, in the last decade, more researchers have begun to extensively investigate the connections between the two disciplines, especially for the decision support systems (Gupta et al., 2022).

In this study, we investigate how the combination of AI and OR (AI–OR) is applied and discussed by the scientific community involved in healthcare. This chapter aims to identify the opportunities for decision-making with combined AI and OR in healthcare, to generate an overview and detect the gaps. We used a systematic way to select the peer-reviewed articles included in this study using MeSH words search in well-known journals in the field. This chapter is structured as follows. In the second section, we present findings that investigate the existing potential of AI-OR applications by highlighting some publications as examples of this contribution worldwide. After that, we focus on the challenges of the implementation of AI-OR solutions in Africa for the healthcare domain.

## 2 AI & OR, Together for Better Healthcare Decision-Making

Decision-making is complex, especially when it involves many stakeholders. Using operations research, quantitative techniques, and data can help present an objective argument. The help that technology has brought to decision-making support methods and systems is critical as it helps increase healthcare efficiency and consistency and reduces clinical errors. With this collaboration between AI and OR, experts' opinions could then be used to fine-tune the quantitative methods. In this section, we classify our findings following the framework presented in Fig. 2.

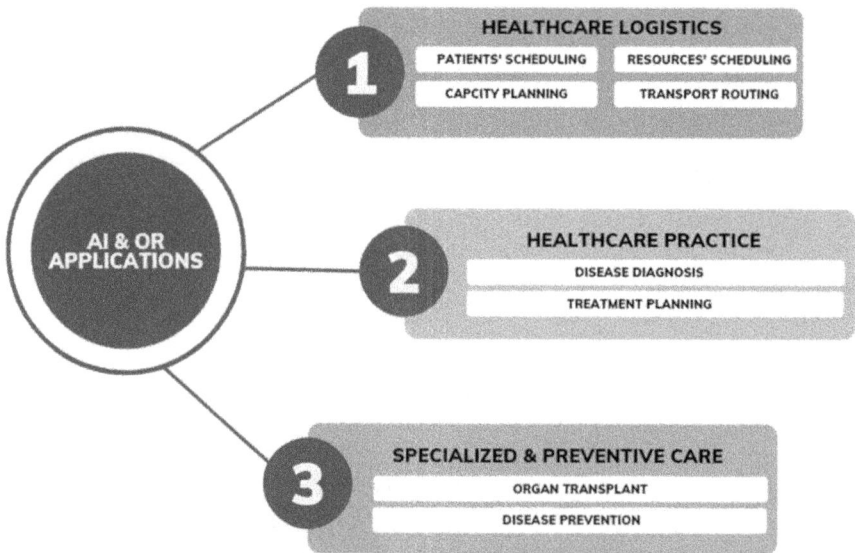

**Fig. 2** The framework of AI–OR applications' classification followed in this study

## 2.1 Logistics and Management for Healthcare

### 2.1.1 Patients' Scheduling

The scheduling of appointments should balance patients' waiting time with clinics' overtime, which is limited by patients' load and staffing. Appointment scheduling systems are used to manage access to service providers and to schedule elective surgeries or lab tests. Many factors impact this decision-making process, such as the experience and input information of the responsible staff, preferences of both providers and patients, and service and arrival time variability (Gupta & Denton, 2008). In a variety of clinical settings, OR researchers use queuing theory and discrete event simulation to propose various appointment strategies (Bailey, 1952; Cayirli et al., 2006). As a result of the complexity of such appointment strategies, the current systems of reporting waiting times rely largely on rolling averages and median estimates, which have limited accuracy (Gupta & Denton, 2008). Machine Learning (ML) algorithms were applied to scheduling optimization problems based on OR models in order to provide evidence of the improvement in waiting time predictions from low acuity Emergency Departments (ED). The ED's waiting time prediction improved by a significant margin using a large set of queuing and service flow variables. The results significantly improved waiting time forecasts which helped in more accurate patients' scheduling in emergency departments (Pak et al., 2021). Other efforts to improve waiting time based on queuing models and AI algorithms helped improve the priority queuing models for Intensive Care Units (ICU) for patients with high Length Of Stay (LOS) in hospitals (Hagen et al., 2013).

Post-Anesthetic Care Unit (PACU), where patients recover after their surgical procedures, also faces scheduling issues. In the event that the PACU fills up, the patients must wait until space becomes available in the operating room, leading to delays and potentially the cancelation of subsequent operating room procedures. Stanford researchers developed algorithms that optimize and learn from operating room procedures to minimize delays caused by PACU unavailability using data from Lucile Packard Children's Hospital Stanford. Additionally, they created two integer programming models to schedule procedures in operating rooms in order to minimize maximum PACU occupancy using ML to predict the required PACU time for each type of surgical procedure and compared the optimized schedule for patients to the existing schedule to validate the improvements (Fairley et al., 2019).

### 2.1.2   Other Resources' Scheduling

The same problem of ED overcrowding can be due to other problems, like lack of resources such as beds, physicians, nurses, and staff, to quote just a few (Pines et al., 2007). These issues compromise the individual Quality Of Care (QOC) but also threatens the communities' public in pandemic situations (System, 2006; Kellermann, 2006; Pines et al., 2007). Looking at these important differences in demand and supply emphasizes how important it remains to allocate resources in the ED efficiently (Jones & Evans, 2008). The Institute Of Medicine's (IOM) report highly recommended that ED and hospitals in general benefit from the OR methods to improve how care is given, its timeliness, and the smoothness of the flow within its crowded departments (System, 2006). This can be done by basing the decision support tools that optimize physician resources on the core logic of OR methods (Jones & Evans, 2008). The unpredictable arrival times of patients to EDs complicated configuring an optimal staffing level. However, if the physicians' staff accounts for these patterns appropriately, consistent weekly and daily patterns emerge (Green et al., 2006).

Additional problems occur in bed allocation. With the changing patients' needs over time, the demand for inpatient beds within the hospitals (Howell et al., 2010). Some facilities allocate beds by specialties or by gender, which generally leads to lower utilization of beds or to over crowdedness (Teow, 2009). If bed allocation is not adjusted, the increasing inconsistency can lead to overflow, which leads to a degradation of the care provided. Distance, waiting time, and other complex factors and constraints should be considered while building the resources' scheduling models (Teow, 2009). These strategies of scheduling should be reviewed and revisited periodically to account for the dynamic changes over time (Teow & Tan, 2008; Cochran & Bharti, 2006). Several tools based on AI to forecast the daily bed need for admissions were applied to help solve these issues (Toerper et al., 2016).

### 2.1.3 Capacity Planning

It is harder for a hospital or a unit to plan its capacity and forecast it in crisis situations like the COVID-19 one. This pandemic poses threats to overwhelm the healthcare systems with unexpected demands for the available resources (Goic et al., 2021). The major issue that both frontline clinicians and policymakers faced was planning and allocating scarce resources such as beds in ICU, staff, and available functioning equipment (Bedford et al., 2020). Managing these demands could not be effectively conducted without a nationwide collective effort that relies on data to forecast the real demands on both individual-level and national basis (Bedford et al., 2020). Re-allocating ventilators and alleviating shortages, many hospitals' clinical studies in the USA resorted to optimization models together with machine learning predictive models that forecast the COVID-19 impacts (Bertsimas et al., 2021). Also, many hospitals in the UK deployed a system called "Capacity Planning and Analysis System" (CPAS) based on mathematical allocation and AI models to assist the planning processes (NHS, 2020; Qian et al., 2021). Some other applications that try to improve healthcare capacity planning use models that are based on approximate reasoning using techniques based on fuzzy sets and logic. They capture uncertainties from the environment and expertise of management using natural language expressions. These models use sophisticated OR model analyses and expert experience knowledge from real-world cases (Turksen, 1992). Such models and others help in the decision-making process to obtain a more effective capacity planning strategy (Abad et al., 2020).

### 2.1.4 Transport Routing

Ambulances are usually considered a scarce resource in healthcare. In disaster response scenarios or pandemic situations, many patients require simultaneous medical aid, and its need remains important. At high frequency, routing models are designed to cope with the uncertainties that are caused by such dynamic situations. They can minimize, among waiting people, the latest service completion time (Talarico et al., 2015). More sophisticated optimization approaches are coordinating between the routing models and the scheduling models in emergency units and hospitals to provide solutions that assess the capacity constraints of the hospitals (Almehdawe et al., 2016). AI, then, interferes with assisting OR in solving such problems using ML algorithms (Yu et al., 2021). Some systems like ALTO, which is based on routing algorithms, can support vehicle routing experts to develop new heuristic algorithms using OR and AI together within a friendly-user interactive graphic system (Potvin et al., 1992).

## 2.2 Healthcare Practice Support

### 2.2.1 Disease Diagnosis

There is no doubt that many diseases can, unfortunately, lead to serious health conditions that can result in death. If the main symptoms of some diseases could be predicted by physicians in advance, these diseases could be handled earlier. This prediction is sometimes hard and time-consuming, which drives researchers to resort to other solutions like data mining and machine learning for medical diagnosis (Tan et al., 2016; Książek et al., 2019; Abdar et al., 2019) and uncertainty estimation (Mohammadzadeh & Zhang, 2019). The US Food and Drug Administration approved in the last two decades many platforms that support the medical system for disease diagnosis based on AI and deep learning (e.g., brain bleed diagnosis with CT, breast density via mammography, paramedic stroke diagnosis, arterial fibrillation detection, etc.) (Topol, 2019; Burlacu et al., 2020). Many efforts have been made since then invested in deploying new solutions that support diagnosis decision-making (Weltz et al., 2022; Singh et al., 2022). For example, clinicians often use angiography to diagnose Coronary Artery Disease (CAD) (Tsipouras et al., 2008; Kim et al., 2015), which is a very expensive procedure that causes many unelectable side effects (Das et al., 2009; Tsipouras et al., 2008). Thus, they replace it with highly accurate data mining-based alternatives. Machine learning algorithms, namely Emotional Neural Networks (EmNNs), are one of those alternatives, combining Particle Swarm Optimization (PSO) with machine learning to diagnose CAD (Shahid & Singh, 2020). Many other examples exist of applications that use high dimensional optimization models to diagnose diseases without noise and uncertainties, like discrete support vector machine predictive models (Lee & Wu, 2009). These applications include discriminate analysis of motility and morphology data in cancer, as well as differentiation of erythemato-squamous diseases and predicting heart disease (Feltus et al., 2006); tumor metastasis; prediction of protein localization sites; pattern recognition of satellite images in the classification of soil types (Lee et al., 2000; Luedi et al., 2005; Mangasarian et al., 1995; Raz & Ben-Ze'ev, 1987); identification of tumor shape and volume in the treatment of sarcoma; multistage discriminate analysis of biomarkers for prediction of early atherosclerosis (Lee et al., 2007); fingerprinting of native and angiogenic microvascular networks for early diagnosis of diabetes, aging, macular degeneracy, and prediction of ultrasonic cell disruption for drug delivery (Lee et al., 2004). The model is accurate in predicting the medical diagnosis, monitoring, and decision-making outcomes in all three areas, ranging from 70 to 100%. This motivates its further application for medical diagnostics, monitoring, and decision-making (Lee & Wu, 2009). The deep breath, a deep learning algorithm used to analyze lung sounds to diagnose and stratify risk for COVID-19 (Glangetas et al., 2021), and the Fuzzy Antenna Optimization method used to diagnose Diabetes disease (Thirugnanam et al., 2016) are two other applications worth mentioning in this challenging field.

### 2.2.2 Treatment Planning

Treatment planning is of great importance, and its optimization has been a challenge for health care providers. Using Automatic Treatment Planning (ATP) tools, patient treatment planning has improved using artificial intelligence and deep learning theories thanks to the rapid growth of computational performance (Poole et al., 1998). Among the most important applications, we can cite the time-consuming, laborious radiotherapy (Wang et al., 2019). The optimization of multicriteria decisions can be accomplished through a variety of algorithms based on rule implementation and reasoning, modeling of prior knowledge, and multicriteria optimization (Sahiner et al., 2019). Rather than automating dissymmetric trade-offs, these algorithms automate their planning and optimization. They have been found to improve the quality and consistency of dose measurements (Russell & Norvig, 2002; Feng et al., 2018). A similar application in treatment planning for cancer patients could be found in (Netherton et al., 2021). Besides, systems like NEWCHEM and ONCOCIN use AI to help support decision-making about new oncology therapies. To guide experimentation and design new optimal protocols, these systems take into account the latest advances in molecular and cellular biology and the cellular-drug interaction (Ardizzone et al., 1988; Shwe et al., 1989). Some other clinical-based decision systems are used for pain management. Pain Management Advisor (PMA), for example, uses AI to optimize plans that help in treatment planning based on pain management algorithms (Shortliffe, 1987; Carson et al., 2013; Wiharto, 2018; Knab et al., 2001). Furthermore, decisions regarding rehabilitation interventions decisions are managed with algorithms of optimization implemented with AI techniques. Another example is the so-called Repeated Incremental Pruning to Produce Error Reduction (RIPPER) algorithm, which is implemented within the decision support system of a Canadian hospital to help physicians optimize decision-making about rehabilitation interventions using machine learning algorithms (Gross et al., 2013). On the other hand, it is well-known that planning treatment for Supraventricular tachyarrhythmia (SVT) in elderly patients is complex (Braunwald, 1988). In order to properly manage SVT in the elderly, it is necessary to make a correct diagnosis, identify underlying causes, and understand the therapeutic options available (Messerli, 2012). SVT, possible therapeutic options, related indications, and contra-indications, along with possible therapeutic options, are all stored in an interactive program, called TACHY. Thus, every physician can find therapeutic recommendations and learn about potential side effects using TACHY (Wang et al., 1998; Messerli, 2012; Rich, 1997; Winston, 1992; KLIX, 1983).

## 2.3 Specialized and Preventive Care

### 2.3.1 Organ Transplant

Organ transplantation is a process that involves a set of input and output factors such as organ allocation, precision treatments, therapeutic algorithms, diagnostic proposals, and results prediction (Briceño, 2020; Sucher & Sucher, 2020). According to traditional methods of organ allocation, the first patient on the waiting list was usually given an organ regardless of the characteristics of the donor and/or recipient (Cruz-Ramirez et al., 2013). Using AI and OR makes it possible to solve problems related to determining whether to include or exclude organ transplant candidates as well as match donors with recipients (Briceño et al., 2020; Ivanics et al., 2020). In the liver transplant case, for example, the demand for transplants is far greater than the supply of deceased donors' organs, so the decision to list and allocate organs is based upon utility maximization (Neuberger, 2016; Kim et al., 2015). Thus, a paradigm based on outcomes that factor in many complex variables (donors, recipients, providers of health care) is necessary to achieve the best results (Busuttil & Tanaka, 2003; Tector et al., 2006). In order to predict the best outcomes with liver organ allocation, different systems have been used to accommodate all the factors, including the scoring systems that are used to support the decision-making processes in the actual healthcare setting, such as the Model for End-Stage Liver Disease (MELD), Balance of Risk (BAR), and Survival Outcome Following Liver Transplantation (SOFT) (Wingfield et al., 2020; Flores & Asrani, 2017).

### 2.3.2   Disease Prevention

Nowadays, patients are generally benefiting from advances in disease prevention and treatment. There is, however, some difficulty leveraging these advances in decision-making because of the uncertainties associated with disease risks and treatment outcomes, as well as the complex nature of the technologies (Kong & Zhang, 2018). Artificial intelligence algorithms are now being used to predict the development of septic shock (Saria, 2014) to diagnose patients with chronic obstructive pulmonary disease, identify and treat patients with schizophrenia, as well as to make many other expert decisions (Anakal & Sandhya, 2017). Additionally, they might be useful in allowing clinicians to assess a patient's health based on a large number of past cases that would historically have been difficult to incorporate into clinical decision-making processes. Using sequential decision-making, an AI framework can be used to recommend alternate treatment paths, detect a patient's health status when measurements are unavailable, and improve treatment and prevention plans as new information becomes available (Reddy et al., 2019). The decision-aid tool is developed using optimization models for infectious Disease Prevention and Epidemic Control (DPEC) (Deng et al., 2013). Several researchers have conducted interesting studies related to disease prevention, such as modeling disease transmission dynamics (Rao & Kakehashi, 2004), allocating resources to prevent outbreaks (Becker & Starczak, 1997; Deng et al., 2013; Wu et al., 2005), and using patient screening to help predict disease incidence (Lee & Pierskalla, 1988; Schwartz et al., 1990; Wein & Zenios, 1996), to quote just a few. As DPEC models become increasingly complex, both optimization techniques and AI algorithms are used to solve these challenging problems (Deng et al., 2013; Marcus et al., 2020).

## 3 Challenges and Opportunities of AI Implementation to Solve Optimization Problems in Africa

### 3.1 State of Art and Opportunities

Historically, Medical AI (MAI), successfully implemented in the USA, was introduced in Africa in the mid-1980s (Hunter et al., 2012). Precisely, MAI deployed in Kenya has improved the health worker–patient interaction quality with evidence of the increased number of symptoms elicited (Hunter et al., 2012). To improve the detection of common and potentially blinding eye disorders, MAI was piloted in Egypt sometime in 1986 (Kastner et al., 1984). Similarly, in Gambia, a probabilistic decision-making system assisted rural health workers in identifying life-threatening conditions in outpatient clinics. Precisely, 88% of cases were detected by men of MAI. Using a cost-and-effectiveness algorithm, computerized treatment aids have also been used by nurses to prescribe drugs in South Africa (Byass, 1987). There have been only a handful of pilot cases in recent times. In South Africa, a multinomial logistic classifier-based system is used in human resource planning in order to predict the length of stay of health workers (Moyo et al., 2018). Another application in resource scheduling and planning is under a partnership between a social enterprise and a researchers' team. Community Health Workers (CHW) schedules are optimized through the yielding program (Brunskill & Lesh, 2010). For disease detection through preventive prediction and diagnosis, many examples in Nigeria, Zambia, and other Eastern countries in Africa showed significantly promising results (Breuninger et al., 2014; Melendez et al., 2017; Bellemo et al., 2019; Onu et al., 2019; Mathenge, 2019; Marcus et al., 2020).

Since the turn of the century, AI has become increasingly important in African healthcare systems. Eliminating medical inefficiencies like misdiagnosis, shortages of doctors, and prolonged waits and recoveries, inefficiency can be eliminated. Over the past century, it has increased significantly. In other healthcare systems, it has proven successful; however, ethical, social, and legal challenges are associated with its implementation. Artificial intelligence may be thought to have no place in African environments (Mahomed, 2018). Therefore, it is essential that African countries guard against the issues and challenges associated with the implementation of AI (Sallstrom et al., 2019).

### 3.2 Challenges

#### 3.2.1 Inadequate Infrastructure

A recent study was conducted by Antwi et al. in 2021 to investigate the African clinicians' concerns regarding AI (Antwi et al., 2021). The yielding results have shown that participants raised their concern about the poor infrastructure in Africa.

Precisely, poor equipment maintenance makes it difficult to them to see the possibility of a sustainable AI for Africa. Actually, the use of AI in the West Africa sub-region is challenging because of poor infrastructure preservation, unreliable system quality management, and outdated medical equipment, among other problems (Antwi et al., 2021; Botwe et al., 2020b; Botwe et al., 2020a). Another major infrastructural inadequacy is the low Internet penetration (39% in 2019) (Owoyemi et al., 2020). Another study conducted in 2019 revealed that nearly half of Africa's population had no access to electricity, which impeded the implementation and execution of digital health initiatives ("More than Half of Sub-Saharan Africans Lack Access to Electricity," 2020).

### 3.2.2 Adoption Is Costly

In developed countries, the cost of assembling an AI-based healthcare solution is difficult to estimate; a study published in 2019 found that it can range anywhere from $6000 to millions of dollars for simple chatbots (*How Much Does Artificial Intelligence (AI) Cost in* 2019?). Another concern raised by Antwi et al.'s study showed that AI cost is a point that doctors doubt (Antwi et al., 2021). Underdeveloped countries may find the cost of implementation and maintenance prohibitive, putting them further behind when it comes to improving their healthcare systems (Ongena et al., 2020; Oren et al., 2020; Meskó et al., 2018; Antwi et al., 2021).

### 3.2.3 Data Is an Issue

Obviously, African countries have low levels of digitalization and use of digital health, which makes it difficult for specialists to find useful data for their models (Akanbi et al., 2012). Thus, AI is a growing field in Africa, but available information is anecdotal, and there is a lack of data on these topics (IDRC, 2019). Most of the datasets that are available and used globally are related to people who are different from African populations physiologically and to systems that are in huge different from the African healthcare systems (Bellemo et al., 2019). Other biases can also come from non-considering that the systems have low-resource settings (Wahl et al., 2018; Mahomed, 2018). In addition, a lack of data could undoubtedly bias the results. For example, during the Covid-19 pandemic, the absence of data was interpreted at first as the absence of cases, but it was because of a lack of surveillance (McCall, 2020).

### 3.2.4 Legal & Policy Issues

African governments have been deploying AI technologies for a while, but policy responses are still at an early stage (Gwagwa et al., 2020). There is still a lack of digital health strategies and policies in some African countries which can guide the monitoring of digital health strategies and the implementation of AI-based solutions (Owoyemi et al., 2020). As an added problem, only 17 of the 55 Member States of the African Union (AU) have comprehensive data protection and privacy legislation, according to Onuoha's 2019 Global Information Society Watch (R, 2019). These privacy issues may arise for users from the AI tolls implemented to help them facilitate some processes and services. In order for AI in Africa to be successful, there is a need for African health professionals, AI manufacturers, policymakers, and all stakeholders to address end users' concerns about AI in several care fields, such as medical imaging (Antwi et al., 2021). Additionally, African policy needs to be addressed at the national, regional, continental, and international levels. All of the stakeholders and actors involved on the continent need to be in a position to reap AI's benefits, as well as mitigate its threats, as a result of government responses (Gwagwa et al., 2020). Moreover, in Africa, it may be difficult for patients and users to understand who is responsible for adverse outcomes of AI in healthcare due to a lack of clarity regarding the identity of the responsible parties (Mahomed, 2018).

## 4   Conclusion

The combination of AI and OR tools to tackle healthcare issues is showing signs of success across the globe. However, despite the growing body of literature and the emerging technologies designed to drive Africa's digital transformation, there are still major challenges for OR models to be widely accepted and used as part of main stream decision-making by clinicians, health managers, and policymakers, especially when solved with AI solutions. This reluctance is mainly due to ethical, financial, legal, and social considerations. In fact, Africa's healthcare systems face several structural challenges. In addition to barriers to accessibility (e.g., rural vs. urban disparities), lack of awareness about health issues can be a barrier to seeking more effective public health policies. Notwithstanding the availability of skilled staff and qualified professionals, affordability puts AI-OR services beyond the reach of African healthcare systems.

## References

Abad, Z. S. H., Maslove, D. M., & Lee, J. (2020). Predicting discharge destination of critically ill patients using machine learning. *IEEE Journal of Biomedical and Health Informatics, 25*(3), 827–837.

Abdar, M., Wijayaningrum, V. N., Hussain, S., Alizadehsani, R., Plawiak, P., Acharya, U. R., & Makarenkov, V. (2019). IAPSO-AIRS: A novel improved machine learning-based system for wart disease treatment. *Journal of Medical Systems, 43*(7), 1–23.

Akanbi, M. O., Ocheke, A. N., Agaba, P. A., Daniyam, C. A., Agaba, E. I., Okeke, E. N., & Ukoli, C. O. (2012). Use of electronic health records in sub-Saharan Africa: Progress and challenges. *Journal of Medicine in the Tropics, 14*(1), 1.

Almehdawe, E., Jewkes, B., & He, Q.-M. (2016). Analysis and optimization of an ambulance offload delay and allocation problem. *Omega, 65*, 148–158. https://doi.org/10.1016/j.omega.2016.01.006

Anakal, S., & Sandhya, P. (2017). Clinical decision support system for chronic obstructive pulmonary disease using machine learning techniques. *2017 International Conference on Electrical, Electronics, Communication, Computer, and Optimization Techniques (ICEECCOT)*: IEEE, 1–5.

Antwi, W. K., Akudjedu, T. N., & Botwe, B. O. (2021). Artificial intelligence in medical imaging practice in Africa: A qualitative content analysis study of radiographers' perspectives. *Insights Into Imaging, 12*(1), 80. https://doi.org/10.1186/s13244-021-01028-z

Ardizzone, E., Bonadonna, F., Gaglio, S., Marcenò, R., Nicolini, C., Ruggiero, C., & Sorbello, F. (1988). Artificial intelligence techniques for cancer treatment planning. *Medical Informatics, 13*(3), 199–210.

Bailey, N. T. (1952). A study of queues and appointment systems in hospital out-patient departments, with special reference to waiting-times. *Journal of the Royal Statistical Society: Series B (Methodological), 14*(2), 185–199.

Becker, N. G., & Starczak, D. N. (1997). Optimal vaccination strategies for a community of households. *Mathematical Biosciences, 139*(2), 117–132.

Bedford, J., Enria, D., Giesecke, J., Heymann, D. L., Ihekweazu, C., Kobinger, G., Lane, H. C., Memish, Z., Oh, M.-D., & Schuchat, A. (2020). COVID-19: Towards controlling of a pandemic. *The Lancet, 395*(10229), 1015–1018.

Bellemo, V., Lim, Z. W., Lim, G., Nguyen, Q. D., Xie, Y., Yip, M. Y., Hamzah, H., Ho, J., Lee, X. Q., & Hsu, W. (2019). Artificial intelligence using deep learning to screen for referable and vision-threatening diabetic retinopathy in Africa: A clinical validation study. *The Lancet Digital Health, 1*(1), e35–e44.

Bertsimas, D., Boussioux, L., Cory-Wright, R., Delarue, A., Digalakis, V., Jacquillat, A., Kitane, D. L., Lukin, G., Li, M., & Mingardi, L. (2021). From predictions to prescriptions: A data-driven response to COVID-19. *Health Care Management Science*, 1–20.

Botwe, B., Schandorf, C., Inkoom, S., & Faanu, A. (2020a). An investigation into the infrastructure and management of computerized tomography units in Ghana. *Journal of Medical Imaging and Radiation Sciences, 51*(1), 165–172.

Botwe, B., Schandorf, C., Inkoom, S., & Faanu, A. (2020b). Status of quality management systems in computed tomography facilities in Ghana. *Radiologic Technology, 91*(4), 324–332.

Braunwald, E. (1988). *Heart disease*.

Breuninger, M., van Ginneken, B., Philipsen, R. H., Mhimbira, F., Hella, J. J., Lwilla, F., van den Hombergh, J., Ross, A., Jugheli, L., & Wagner, D. (2014). Diagnostic accuracy of computer-aided detection of pulmonary tuberculosis in chest radiographs: A validation study from sub-Saharan Africa. *PLoS One, 9*(9), e106381.

Briceño, J. (2020). Artificial intelligence and organ transplantation: Challenges and expectations. *Current Opinion in Organ Transplantation, 25*(4), 393–398. https://doi.org/10.1097/mot.0000000000000775

Briceño, J., Ayllón, M. D., & Ciria, R. (2020). Machine-learning algorithms for predicting results in liver transplantation: The problem of donor-recipient matching. *Current Opinion in Organ Transplantation, 25*(4), 406–411. https://doi.org/10.1097/mot.0000000000000781

Brunskill, E., & Lesh, N. (2010). Routing for rural health: optimizing community health worker visit schedules. *2010 AAAI Spring Symposium Series*.

Burlacu, A., Iftene, A., Jugrin, D., Popa, I. V., Lupu, P. M., Vlad, C., & Covic, A. (2020). Using artificial intelligence resources in dialysis and kidney transplant patients: A literature review. *BioMed Research International, 2020*, 9867872. https://doi.org/10.1155/2020/9867872

Busuttil, R. W., & Tanaka, K. (2003). The utility of marginal donors in liver transplantation. *Liver Transplantation, 9*(7), 651–663. https://doi.org/10.1053/jlts.2003.50105

Byass, P. (1987). Computers in Africa: Appropriate technology? *The Computer Bulletin, 3*(2), 17–17.

Capan, M., Khojandi, A., Denton, B. T., Williams, K. D., Ayer, T., Chhatwal, J., Kurt, M., Lobo, J. M., Roberts, M. S., Zaric, G., Zhang, S., & Schwartz, J. S. (2017). From data to improved decisions: Operations research in healthcare delivery. *Medical Decision Making, 37*(8), 849–859. https://doi.org/10.1177/0272989x17705636

Carson, E., Deutsch, T., & Ludwig, E. (2013). *Dealing with medical knowledge: Computers in clinical decision making.* Springer Science & Business Media.

Cayirli, T., Veral, E., & Rosen, H. (2006). Designing appointment scheduling systems for ambulatory care services. *Health Care Management Science, 9*(1), 47–58.

Cochran, J. K., & Bharti, A. (2006). Stochastic bed balancing of an obstetrics hospital. *Health Care Management Science, 9*(1), 31–45.

Cruz-Ramirez, M., Hervas-Martinez, C., Fernandez, J. C., Briceno, J., & de la Mata, M. (2013). Predicting patient survival after liver transplantation using evolutionary multi-objective artificial neural networks. *Artificial Intelligence in Medicine, 58*(1), 37–49. https://doi.org/10.1016/j.artmed.2013.02.004

Das, R., Turkoglu, I., & Sengur, A. (2009). Effective diagnosis of heart disease through neural networks ensembles. *Expert Systems with Applications, 36*(4), 7675–7680.

Dean, B. V. (1958). Application of operations research to managerial decision making. *Administrative Science Quarterly, 3*(3), 412–428. https://doi.org/10.2307/2390719

Deng, Y., Shen, S., & Vorobeychik, Y. (2013). Optimization methods for decision making in disease prevention and epidemic control. *Mathematical Biosciences, 246*(1), 213–227.

Dornemann, J., Rückert, N., Fischer, K., & Taraz, A. (2020). Artificial intelligence and operations research in maritime logistics. *Data science in maritime and city logistics: Data-driven solutions for logistics and sustainability. Proceedings of the Hamburg International Conference of Logistics (HICL), Vol. 30*: Berlin: epubli GmbH, 337–381.

Fairley, M., Scheinker, D., & Brandeau, M. L. (2019). Improving the efficiency of the operating room environment with an optimization and machine learning model. *Health Care Management Science, 22*(4), 756–767.

Feltus, F., Lee, E., Costello, J., Plass, C., & Vertino, P. (2006). DNA signatures associated with CpG island methylation states. *Genomics, 87*, 572–579.

Feng, M., Valdes, G., Dixit, N., & Solberg, T. D. (2018). Machine learning in radiation oncology: opportunities, requirements, and needs. *Frontiers in Oncology, 8*, 110.

Flores, A., & Asrani, S. K. (2017). The donor risk index: A decade of experience. *Liver Transplantation, 23*(9), 1216–1225. https://doi.org/10.1002/lt.24799

Gass, S. I. (2003). *Linear programming: Methods and applications.* Courier Corporation.

Glangetas, A., Hartley, M.-A., Cantais, A., Courvoisier, D. S., Rivollet, D., Shama, D. M., Perez, A., Spechbach, H., Trombert, V., & Bourquin, S. (2021). Deep learning diagnostic and risk-stratification pattern detection for COVID-19 in digital lung auscultations: Clinical protocol for a case–control and prospective cohort study. *BMC Pulmonary Medicine, 21*(1), 1–8.

Goic, M., Bozanic-Leal, M. S., Badal, M., & Basso, L. J. (2021). COVID-19: Short-term forecast of ICU beds in times of crisis. *PLoS One, 16*(1), e0245272.

Gomes, C. P. (2000). Artificial intelligence and operations research: Challenges and opportunities in planning and scheduling. *The Knowledge Engineering Review, 15*(1), 1–10. https://doi.org/10.1017/S0269888900001090

Green, L. V., Soares, J., Giglio, J. F., & Green, R. A. (2006). Using queueing theory to increase the effectiveness of emergency department provider staffing. *Academic Emergency Medicine, 13*(1), 61–68.

Gross, D. P., Zhang, J., Steenstra, I., Barnsley, S., Haws, C., Amell, T., McIntosh, G., Cooper, J., & Zaiane, O. (2013). Development of a computer-based clinical decision support tool for selecting appropriate rehabilitation interventions for injured workers. *Journal of Occupational Rehabilitation, 23*(4), 597–609. https://doi.org/10.1007/s10926-013-9430-4

Gupta, D., & Denton, B. (2008). Appointment scheduling in health care: Challenges and opportunities. *IIE Transactions, 40*(9), 800–819.

Gupta, S., Modgil, S., Bhattacharyya, S., & Bose, I. (2022). Artificial intelligence for decision support systems in the field of operations research: Review and future scope of research. *Annals of Operations Research, 308*(1), 215–274. https://doi.org/10.1007/s10479-020-03856-6

Gwagwa, A., Kraemer-Mbula, E., Rizk, N., Rutenberg, I., & De Beer, J. (2020). Artificial intelligence (AI) deployments in Africa: Benefits, challenges and policy dimensions. *The African Journal of Information and Communication, 26*, 1–28.

Hagen, M. S., Jopling, J. K., Buchman, T. G., & Lee, E. K. (2013). Priority queuing models for hospital intensive care units and impacts to severe case patients. *AMIA Annual Symposium Proceedings: American Medical Informatics Association*, 841.

Holsapple, C. W., Jacob, V. S., & Whinston, A. B. (1994). *Operations research and artificial intelligence*. Intellect Books.

*How Much Does Artificial Intelligence (AI) Cost in 2019?*: AI, ML, NLP consulting and software development. 2019. Available from https://azati.ai/how-much-does-it-cost-to-utilize-machine-learning-artificial-intelligence/

Howell, E., Bessman, E., Marshall, R., & Wright, S. (2010). Hospitalist bed management effecting throughput from the emergency department to the intensive care unit. *Journal of Critical Care, 25*(2), 184–189.

Hunter, J., Cookson, J., & Wyatt, J. (2012) *AIME 89: Second European conference on artificial intelligence in medicine, London, August 29th–31st 1989. Proceedings*. Springer Science & Business Media.

IDRC, O. I. (2019). *Government artificial intelligence readiness index 2019*. Available from https://www.oxfordinsights.com/government-ai-readiness-index-2020.

Inozemtsev, V., Ivleva, M., & Ivlev, V. (2017). Artificial intelligence and the problem of computer representation of knowledge. *Proceedings of the, 1151*.

Ivanics, T., Patel, M. S., Erdman, L., & Sapisochin, G. (2020). Artificial intelligence in transplantation (machine-learning classifiers and transplant oncology). *Current Opinion in Organ Transplantation, 25*(4), 426–434. https://doi.org/10.1097/mot.0000000000000773

Jiang, F., Jiang, Y., Zhi, H., Dong, Y., Li, H., Ma, S., Wang, Y., Dong, Q., Shen, H., & Wang, Y. (2017). Artificial intelligence in healthcare: Past, present and future. *Stroke and Vascular Neurology, 2*(4), 230–243. https://doi.org/10.1136/svn-2017-000101

Jones, S. S., & Evans, R. S. (2008). An agent based simulation tool for scheduling emergency department physicians. *AMIA Annual Symposium Proceedings: American Medical Informatics Association*, 338.

Kastner, J. K., Dawson, C. R., Weiss, S. M., Kern, K. B., & Kulikowski, C. A. (1984). An expert consultation system for frontline health workers in primary eye care. *Journal of Medical Systems, 8*(5), 389–397.

Kellermann, A. L. (2006). Crisis in the emergency department. *The New England Journal of Medicine, 355*(13), 1300.

Kim, H. Y., Kim, E. K., Kim, S. M., Song, Y. B., Hahn, J.-Y., Choi, S.-H., Gwon, H.-C., Lee, S. H., Choe, Y. H., & Oh, J. K. (2015). Fractional myocardial mass: A new index for diagnosis and treatment of coronary artery disease. *Journal of the American College of Cardiology, 65*(10S), A1269–A1269.

Kim, W. R., Lake, J. R., Smith, J. M., Skeans, M. A., Schladt, D. P., Edwards, E. B., Harper, A. M., Wainright, J. L., Snyder, J. J., Israni, A. K., & Kasiske, B. L. (2015). OPTN/SRTR 2013 Annual Data Report: Liver. *American Journal of Transplantation, 15*(Suppl 2), 1–28. https://doi.org/10.1111/ajt.13197

Klix, F. (1983). Principles of artificial-intelligence-Nilsson, NJ. Johann Ambrosius Barth Verlag Im Weiher 10, D-69121 Heidelberg, Germany.

Knab, J. H., Wallace, M. S., Wagner, R. L., Tsoukatos, J., & Weinger, M. B. (2001). The use of a computer-based decision support system facilitates primary care physicians' management of chronic pain. *Anesthesia & Analgesia, 93*(3), 712–720.

Kobbacy, K. A., Vadera, S., & Rasmy, M. H. (2007). AI and OR in management of operations: History and trends. *Journal of the Operational Research Society, 58*(1), 10–28.

Kong, N., & Zhang, S. (2018). *Decision analytics and optimization in disease prevention and treatment*. Wiley.

Książek, W., Abdar, M., Acharya, U. R., & Pławiak, P. (2019). A novel machine learning approach for early detection of hepatocellular carcinoma patients. *Cognitive Systems Research, 54,* 116–127.

Kusiak, A. (1987). Artificial intelligence and operations research in flexible manufacturing systems. *INFOR: Information Systems and Operational Research, 25*(1), 2–12.

Lee, E., Ashfaq, S., Jones, D., Rhodes, S., Weintrau, W., Hopper, C., Vaccarino, V., Harrison, D., & Quyyumi, A. (2007). *Prediction of early atherosclerosis in healthy adults via novel markers of oxidative stress and d-ROMs*: Working paper.

Lee, E. K., Gallagher, R. I., Campbell, A. M., & Prausnitz, M. R. (2004) Statistical analysis of acoustic spectra. *IEEE Transactions on Biomedical Engineering, 51*(1).

Lee, Y.-J., Mangasarian, O., & Wolberg, W. (2000). Breast cancer survival and chemotherapy: A support vector machine analysis. *DIMACS Series in Discrete Mathematics and Theoretical Computer Science, 55,* 1–10.

Lee, H. L., & Pierskalla, W. P. (1988). Mass screening models for contagious diseases with no latent period. *Operations Research, 36*(6), 917–928.

Lee, E. K., & Wu, T.-L. (2009). *Disease diagnosis: Optimization-based methods*.

Luedi, P. P., Hartemink, A. J., & Jirtle, R. L. (2005). Genome-wide prediction of imprinted murine genes. *Genome Research, 15*(6), 875–884.

Mahomed, S. (2018). Healthcare, artificial intelligence and the Fourth Industrial Revolution: Ethical, social and legal considerations. *South African Journal of Bioethics and Law, 11*(2), 93–95.

Mangasarian, O. L., Street, W. N., & Wolberg, W. H. (1995). Breast cancer diagnosis and prognosis via linear programming. *Operations Research, 43*(4), 570–577.

Marcus, J. L., Sewell, W. C., Balzer, L. B., & Krakower, D. S. (2020). Artificial intelligence and machine learning for HIV prevention: Emerging approaches to ending the epidemic. *Current HIV/AIDS Reports, 17*(3), 171–179. https://doi.org/10.1007/s11904-020-00490-6

Mathenge, W. C. (2019). Artificial intelligence for diabetic retinopathy screening in Africa. *The Lancet Digital Health, 1*(1), e6–e7.

McCall, B. (2020). COVID-19 and artificial intelligence: Protecting health-care workers and curbing the spread. *The Lancet Digital Health, 2*(4), e166–e167.

Melendez, J., Philipsen, R., Chanda-Kapata, P., Sunkutu, V., Kapata, N., & van Ginneken, B. (2017). Automatic versus human reading of chest X-rays in the Zambia National Tuberculosis Prevalence Survey. *The International Journal of Tuberculosis and Lung Disease, 21*(8), 880–886.

Meskó, B., Hetényi, G., & Győrffy, Z. (2018). Will artificial intelligence solve the human resource crisis in healthcare? *BMC Health Services Research, 18*(1), 1–4.

Messerli, F. H. (2012). *Cardiovascular disease in the elderly*. Springer Science & Business Media.

Mohammadzadeh, A., & Zhang, W. (2019). Dynamic programming strategy based on a type-2 fuzzy wavelet neural network. *Nonlinear Dynamics, 95*(2), 1661–1672.

More than Half of Sub-Saharan Africans Lack Access to Electricity. (2020). *The Economist*.

Morse, P. M., Kimball, G. E., & Gass, S. I. (2003). *Methods of operations research*. Courier Corporation.

Moyo, S., Doan, T. N., Yun, J. A., & Tshuma, N. (2018). Application of machine learning models in predicting length of stay among healthcare workers in underserved communities in South Africa. *Human Resources for Health, 16*(1), 1–9.

Netherton, T. J., Cardenas, C. E., Rhee, D. J., & Beadle, B. M. (2021). The emergence of artificial intelligence within radiation oncology treatment planning. *Oncology, 99*(2), 124–134.

Neuberger, J. (2016). Liver transplantation in the United Kingdom. *Liver Transplantation, 22*(8), 1129–1135. https://doi.org/10.1002/lt.24462

NHS. (2020). Trials begin of machine learning system to help hospitals plan and manage COVID-19 treatment resources developed by NHS Digital and University of Cambridge.

Ongena, Y. P., Haan, M., Yakar, D., & Kwee, T. C. (2020). Patients' views on the implementation of artificial intelligence in radiology: Development and validation of a standardized question-naire. *European Radiology, 30*(2), 1033–1040.

Onu, C. C., Lebensold, J., Hamilton, W. L., & Precup, D. (2019). *Neural transfer learning for cry-based diagnosis of perinatal asphyxia.* arXiv preprint arXiv:1906.10199.

Oren, O., Gersh, B. J., & Bhatt, D. L. (2020). Artificial intelligence in medical imaging: Switching from radiographic pathological data to clinically meaningful endpoints. *The Lancet Digital Health, 2*(9), e486–e488.

Owoyemi, A., Owoyemi, J., Osiyemi, A., & Boyd, A. (2020). Artificial intelligence for healthcare in Africa. *Frontiers in Digital Health, 2*, 6.

Pak, A., Gannon, B., & Staib, A. (2021). Predicting waiting time to treatment for emergency department patients. *International Journal of Medical Informatics, 145*, 104303.

Pines, J. M., Garson, C., Baxt, W. G., Rhodes, K. V., Shofer, F. S., & Hollander, J. E. (2007). ED crowding is associated with variable perceptions of care compromise. *Academic Emergency Medicine, 14*(12), 1176–1181.

Poole, D., Mackworth, A., & Goebel, R. (1998). *Computational intelligence.*

Potvin, J.-Y., Lapalme, G., & Rousseau, J.-M. (1992). Integration of AI and OR techniques for computer-aided algorithmic design in the vehicle routing domain. *Artificial Intelligence in Operational Research*: Springer, pp. 205–213.

Qian, Z., Alaa, A. M., & van der Schaar, M. (2021). CPAS: The UK's national machine learning-based hospital capacity planning system for COVID-19. *Machine Learning, 110*(1), 15–35.

R, O. (2019). *AI in Africa: Regional data protection and privacy policy harmonisation. In Association for Progressive Communications (APC), Article 19, & Swedish International Development Cooperation Agency (Sida) (Eds.), Global information society watch 2019:Artificial intelligence: Human rights, social justice and development.* Available at: https://giswatch. org/sites/default/files/gisw2019.

Rais, A., & Viana, A. (2011). Operations research in healthcare: A survey. *International Transactions in Operational Research, 18*(1), 1–31. https://doi.org/10.1111/j.1475-3995.2010. 00767.x

Rao, A. S. S., & Kakehashi, M. (2004). A combination of differential equations and convolution in understanding the spread of an epidemic. *Sadhana, 29*(3), 305.

Raz, A., & Ben-Ze'ev, A. (1987). Cell-contact and-architecture of malignant cells and their relationship to metastasis. *Cancer and Metastasis Reviews, 6*(1), 3–21.

Reddy, S., Fox, J., & Purohit, M. P. (2019). Artificial intelligence-enabled healthcare delivery. *Journal of the Royal Society of Medicine, 112*(1), 22–28. https://doi.org/10.1177/ 0141076818815510

Rich, S. (1997). *Heart disease: A textbook of cardiovascular medicine* (5th ed.), Ed: E. Braunwald. WB Saunders, Philadelphia.

Royston, G. (2016). One hundred years of Operational Research in Health—UK 1948–2048. *Operational Research for Emergency Planning in Healthcare: Volume 2*: Springer, pp. 316–338.

Russell, S., & Norvig, P. (2002). *Artificial intelligence: A modern approach.*

Sahiner, B., Pezeshk, A., Hadjiiski, L. M., Wang, X., Drukker, K., Cha, K. H., Summers, R. M., & Giger, M. L. (2019). Deep learning in medical imaging and radiation therapy. *Medical Physics, 46*(1), e1–e36.

Sallstrom, L., Morris, O., & Mehta, H. (2019). Artificial intelligence in Africa's healthcare: Ethical considerations. *Observer Research Foundation Issue Brief, 4.*

Saria, S. (2014). A $3 Trillion challenge to computational scientists: Transforming healthcare delivery. *IEEE Intelligent Systems, 29*(4), 82–87. https://doi.org/10.1109/MIS.2014.58

Schwartz, J. S., Kinosian, B. P., Pierskalla, W. P., & Lee, H. (1990). Strategies for screening blood for human immunodeficiency virus antibody: Use of a decision support system. *JAMA, 264*(13), 1704–1710.

Shahid, A. H., & Singh, M. (2020). A novel approach for coronary artery disease diagnosis using hybrid particle swarm optimization based emotional neural network. *Biocybernetics and Biomedical Engineering, 40*(4), 1568–1585.

Shortliffe, E. H. (1987). Computer programs to support clinical decision making. *JAMA, 258*(1), 61–66.

Shwe, M., Tu, S., & Fagan, L. (1989). Validating the knowledge base of a therapy planning system. *Methods of Information in Medicine, 28*(01), 36–50.

Singh, N. B., Singh, M. M., & Sarkar, A. (2022). Deep learning architectures, libraries and frameworks in healthcare. *Deep learning, machine learning and IoT in biomedical and health informatics* (pp. 221–248). CRC Press.

Sucher, R., & Sucher, E. (2020). Artificial intelligence is poised to revolutionize human liver allocation and decrease medical costs associated with liver transplantation. *Hepatobiliary Surgery and Nutrition, 9*(5), 679–681. https://doi.org/10.21037/hbsn-20-458

System, I. o. M. C. o. t. F. o. E. C. i. t. U. H. (2006). The future of emergency care in the United States health system. *Annals of Emergency Medicine, 48*(2), 115–120.

Talarico, L., Meisel, F., & Sörensen, K. (2015). Ambulance routing for disaster response with patient groups. *Computers & Operations Research, 56*, 120–133.

Tan, P.-N., Steinbach, M., & Kumar, V. (2016). *Introduction to data mining*. Pearson Education India.

Tector, A. J., Mangus, R. S., Chestovich, P., Vianna, R., Fridell, J. A., Milgrom, M. L., Sanders, C., & Kwo, P. Y. (2006). Use of extended criteria livers decreases wait time for liver transplantation without adversely impacting posttransplant survival. *Annals of Surgery, 244*(3), 439–450. https://doi.org/10.1097/01.sla.0000234896.18207.fa

Teow, K. L. (2009). Practical operations research applications for healthcare managers. *Annals of the Academy of Medicine, 38*(6), 564–573.

Teow, K., & Tan, W. (2008). Allocation of hospital beds in an existing hospital. *Journal of Operations and Logistics, 2*(2).

Thirugnanam, M., Thirugnanam, T., & Swarnalatha, A. (2016). *Using fuzzy ant colony optimization for diagnosis of diabetes disease.*

Toerper, M. F., Flanagan, E., Siddiqui, S., Appelbaum, J., Kasper, E. K., & Levin, S. (2016). Cardiac catheterization laboratory inpatient forecast tool: A prospective evaluation. *Journal of the American Medical Informatics Association, 23*(e1), e49–e57.

Topol, E. J. (2019). High-performance medicine: The convergence of human and artificial intelligence. *Nature Medicine, 25*(1), 44–56. https://doi.org/10.1038/s41591-018-0300-7

Tsipouras, M. G., Exarchos, T. P., Fotiadis, D. I., Kotsia, A. P., Vakalis, K. V., Naka, K. K., & Michalis, L. K. (2008). Automated diagnosis of coronary artery disease based on data mining and fuzzy modeling. *IEEE Transactions on Information Technology in Biomedicine, 12*(4), 447–458.

Turksen, I. (1992). Fuzzy ordinal models for AI—OR. *Artificial Intelligence in Operational Research*, pp. 115–126.

Wahl, B., Cossy-Gantner, A., Germann, S., & Schwalbe, N. R. (2018). Artificial intelligence (AI) and global health: How can AI contribute to health in resource-poor settings? *BMJ Global Health, 3*(4), e000798.

Wang, S., Xie, J., Sada, M., Doherty, T. M., & French, W. J. (1998). TACHY: An expert system for the management of supraventricular tachycardia in the elderly. *American Heart Journal, 135*(1), 82–87.

Wang, C., Zhu, X., Hong, J. C., & Zheng, D. (2019). Artificial intelligence in radiotherapy treatment planning: Present and future. *Technology in Cancer Research & Treatment, 18*, 1533033819873922.

Wein, L. M., & Zenios, S. A. (1996). Pooled testing for HIV screening: Capturing the dilution effect. *Operations Research, 44*(4), 543–569.

Weltz, J., Volfovsky, A., & Laber, E. B. (2022). Reinforcement learning methods in public health. *Clinical Therapeutics*. https://doi.org/10.1016/j.clinthera.2021.11.002.

Wiharto, W. (2018). Clinical decision support systems theory and practice. *Jurnal Teknosains, 7*(2), 148–150.

Wingfield, L. R., Ceresa, C., Thorogood, S., Fleuriot, J., & Knight, S. (2020). Using artificial intelligence for predicting survival of individual grafts in liver transplantation: A systematic review. *Liver Transplantation, 26*(7), 922–934. https://doi.org/10.1002/lt.25772

Winston 3rd, P. H. A. I. (1992). Addison Wesley Publishing Company. New York.

Wu, J. T., Wein, L. M., & Perelson, A. S. (2005). Optimization of influenza vaccine selection. *Operations Research, 53*(3), 456–476.

Yu, K.-H., Beam, A. L., & Kohane, I. S. (2018). Artificial intelligence in healthcare. *Nature Biomedical Engineering, 2*(10), 719–731. https://doi.org/10.1038/s41551-018-0305-z

Yu, M., Kollias, D., Wingate, J., Siriwardena, N., & Kollias, S. (2021). Machine learning for predictive modelling of ambulance calls. *Electronics, 10*(4), 482.

# A CODAS Disaggregation Approach for Urban Rainwater Management

**Mouna Regaieg Cherif and Hela Moalla Frikha**

**Abstract** In the field of Operations Research, there is a growing focus on environmental issues. Water is the major basis of stable development around the world. In addition, water scarcity is one important worldwide problem, especially the rainfall rarity. This in turn places significant constraints for decision makers in optimizing water alternatives, mostly in the case of Tunisian urban communities. Howbeit, choosing the rainwater management technique is not an easy task. Therefore, operational research finds of course its place to solve such a problem. The gap between theory and practice of rainwater management is investigated through integrating preferences with CODAS multi-criteria approach and linear optimization. Thus, we aim in this chapter is to define a CODAS parameter through a series of linear mathematical programs solved by CPLEX software.

## 1 Introduction

Water is very necessary in life. Water scarcity is one of the many water problems that exist today and will become even more visible in the future (Karnib, 2004). Furthermore, the people density, economic development required the correct use of water resources which will receive more attention in the future. In the worldwide, various alternatives water resources in arid and semi-arid regions (groundwater, wastewater treatment, water desalination and rainwater harvesting) that treated to solve water scarcity problems. Really, many countries in North Africa as well as Tunisia dealt with water resources and its applications (Cherif & Frikha, 2022a, 2022b; Elleuch et al., 2019; Ben Amor & Frikha, 2018).

M. R. Cherif (✉)
Modeling and Optimization for Decisional, Industrial and Logistic Systems Laboratory (MODILS) Research Laboratory, Faculty of Economics Sciences and Management of Sfax, University of Sfax, Sfax, Tunisia

H. M. Frikha
Optimisation Logistique et Informatique Décisionnelle (OLID) Research Laboratory, Higher Institute of Industrial Management of Sfax, University of Sfax, Sfax, Tunisia

© The Author(s), under exclusive license to Springer Nature Switzerland AG 2022                    71
H. Masri (ed.), *Africa Case Studies in Operations Research*, Contributions to
Management Science, https://doi.org/10.1007/978-3-031-17008-9_4

In the Operational Research field, interest in water issues is increasing especially in optimization techniques (e.g., multi-criteria decision analysis and mathematical programming). In fact, water resource management is a rather new phenomenon in the operational research literature. In addition, Multi-Criteria Decision Analysis (MCDA) is a valuable problem-solving tool characterized by objectives, criteria, and group decision makers. It is a branch of one of the greatest areas of operational research and treats with finding optimal outcomes in complex scenarios including different indicators, conflicting objectives, and criteria (Kumar et al., 2017).

Several researchers have found that Multi-Criteria Decision Analysis expands an effective tool for managing water resources (Fernandes et al., 1999; Almasri & Kaluarachchi, 2005; Zardari et al., 2015; Shen et al., 2016; Alamanos et al., 2018; Toosi et al., 2020). For instance, Raju et al. (2000) employed five MCDA methods for sustainable water resources ranking in group decision-making environment. Besides, based on sustainable criteria Calizaya et al. (2010) applied Analytical Hierarchy Process (AHP) theory to attain a sustainable strategy for water resources management. Ammar et al. (2016) dealt to categorize, compare and identify four main methodologies for selecting Harvested rainwater locations used in arid and semi-arid regions.

The resolution of a multi-criteria problem requires setting the parameter values for each method. However, the decision maker does not have clear crisp information on these parameters. Due to the imprecision of human thought, personal preferences and judgments cannot be expressed with confidence. Although, the determination of the parameters of the mult-icriteria model is carried out using two means: The first is the direct elicitation (aggregation method) that is to say we ask the DM to fix parameter values, which is more or less difficult for him. The second is indirect elicitation (disaggregation method); we simply start with a set of preference relations to find the parameters. That is, the DM expresses a judgment on a subset of alternatives (for example, $a_1Pa_2$ or the preference intensity of alternative $a_1$ over alternative $a_2$ is greater than that of alternative $a_3$ over alternative $a_4$) or/and a subset of criteria (for example, criterion $C_i$ is more important than criterion $C_j$). Indeed, the preference disaggregation approach in multi-criteria analysis is to determine preference modeling elements from preferential structures provided by the decision maker, taking into account the retained decisional problematic (Dias & Climaco, 2000). Therefore, the disaggregation method deals with the opposite of the aggregation method, as explained in Table 1.

On the one hand, the drawback of mult-icriteria aggregation methods is their sensitivity to the chosen parameters. On the other hand, one crucial problem in

**Table 1** Aggregation versus disaggregation approaches

| Preference model | Aggregation method | Disaggregation method |
|---|---|---|
| Input | Performance matrix & Preferred parameters | Performance matrix & Global preferences |
| Output | Global preferences | Preferred parameters |

MCDA is to assess the parameters, and more precisely, it is often difficult to define threshold parameter values because it is inexact and not clear to the DM.

Hence, the above research failed to trait group preference disaggregation issue within CODAS approach. In this study, we aim to develop a new disaggregation method for eliciting CODAS' threshold parameters based on group preference information given by a group of the urban committee according to their previous experience for rainwater management. In this chapter, we address to minimize the inconsistency between the model obtained with this parameter and the subjective preferences relations defined by the urban manager. Inference procedure is perfect through the application of Operational Research to the water resource management.

In addition to the introduction, the rest of this paper is organized as follows. In Sect. 2, different steps of CODAS method will be presented with a review of its application. The third section will offer a literature review of the preference disaggregation method. The proposed methodological for threshold values elicitation approach from the CODAS method will be explained in Sect. 4. We investigate the power of the decision-making procedure, which comes from the integration of the Operational Research methodologies and the environmental dimension to improve water resources management. Section 5 will handle a numerical example with a sensitivity analysis to discuss the effects of our Operational Research methodologies on the urban rainwater management. Finally, Sect. 6 gives some conclusions and some directions for further research.

## 2 Literature Review

The preference disaggregation approaches are characterized by the inference of the parameter values of MCDA methods. Indeed, the history of the disaggregation approach begins with the UTA method (UTility Additives) developed by Jacquet-Lagreze and Siskos (1982). This method aims to elicit one or more additive value functions from a given ranking using special linear programming techniques. Indeed, the UTADIS method (UTilités Additives DIScriminantes) is a preference disaggregation method in sorting problems (Devaud et al., 1980). Its aim is to elicit additive value functions. Using linear programming, UTA II model (Siskos, 1980) is used to determine partial utilities and criteria weights by disaggregating global preferences. Greco et al. (2008) developed a UTA GMS method, which makes it possible to determine all the additive value functions compatible with the DM's preferences. Within an uncertain environment, Siskos (1983) developed Stochastic UTA. Dimitrios and Nikolaos (2017) employed UTASTAR to support Decision Risk in the marine environment to choose ship-to-ship transfer area.

The PROMETHEE (Brans & Vincke, 1985) and ELECTRE (Roy, 1991) families are the principal outranking methods. Various researcher applied disaggregation procedure to deduce the parameters of the outranking methods. In ELECTRE family methods, Mousseau et al. (2001) suggested a partial inference approach that deduces only the criteria weight values and the cut level for a correct assignment of the alternatives. In addition, Kiss et al. (1994) explained the overall DM's preferences from pairwise comparisons of alternatives. An interactive aggregation/disaggregation procedure is used in the ELECCALC system, which allows DM to evaluate the parameters of ELECTRE II. On the other hand, Frikha and Charfi (2018) inferred ELECTRE I criteria weights using binary outranking relations. In multi-criteria sorting problems and, more precisely, in ELECTRE TRI method, Mousseau and Slowinski (1998) determined all the ELECTRE TRI parameters at the same time from the assignment examples. Although they proposed an interactive aggregation-disaggregation approach that simultaneously infers weights, profiles and threshold parameters. Similarly, the method of An and Mousseau (2002) finds the profiles of ELECTRE TRI, assuming that the other parameters are known. Many techniques have been proposed for the elicitation of the parameters of ELECTRE TRI method for a group DM (Cailloux et al., 2012; Damart et al., 2007; Dias & Climaco, 2000).

Yet, some works on the elicitation of the parameters exist for the PROMETHEE method. To help the DM to rank the alternatives and select the best, Ozerol and Karasakal (2007) proposed three interactive approaches to obtain the parameters for PROMETHEE I and II. Besides, Frikha et al. (2010) evolved a disaggregation approach making it possible to infer the relative weights of the criteria in PROMETHEE II from the partial data given by the decision maker. Furthermore, to define both preference and indifference thresholds in PROMETHEE II method, Frikha et al. (2011) developed an interactive disaggregation approach based on some preference relations. In the same context of preferences disaggregation, Zhaoxu and Min (2010) have proposed a similar approach which is also limited to determine by linear programming the PROMETHEE criteria weight parameters indirectly. There-fore, Frikha et al. (2017) extended a new preference disaggregation approach reducing subjectivity in determining objective values for the criteria weights as well the two parameters (preference and indifference) in the PROMETHEE method from binary preference relations supplied by the decision maker. Besides, in the Data Envelopment Analysis (DEA), Jahanshahloo et al. (2011) obtained the common weights based on the DM's preference information. Li and Tzeng (2009) used the maximum mean de-entropy (MMDE) algorithm to identify the threshold value of the DEMATEL method. While Labreuche et al. (2015) suggested an interactive elicitation process of the weights of an Ordered Weighted Average OWA aggregation function of the MACBETH approach from pairwise comparisons between binary alternatives given by the DM. Also, Ghram and Frikha (2018) determined the criteria relative importance coefficients within the ARAS method based on preference relations defined by the DM with some comparison criteria weights.

## 3   The CODAS Method

CODAS (COmbinative Distance-based ASsessment) method developed by Mehdi Keshavarz Ghorabaee et al. (2016) is one of the newest MCDA methods. Like Weighted Aggregated Sum Product ASsessment (WASPAS) (Zavadskas et al.,

2012; Chakraborty & Zavadskas, 2014) method, COmplex PRoportional ASsessment (COPRAS) (Zavadskas et al., 1994) method, Technique for Ordering Preference by Similarity to Ideal Solution (TOPSIS) (Hwang & Yoon, 1981) method, Evaluation based on Distance from Average Solution (EDAS) (Keshavarz Ghorabaee et al., 2015) method and VIekriterijumsko KOmpromisno Rangiranje (VIKOR) (Opricovic & Tzeng, 2004) method, CODAS is a distance-based method. The ranking of alternatives is determined using two measures. The main and primary measure of assessment is related to the Euclidean distance of alternatives from the "Negative-Ideal Solution." The secondary measure is the Taxicab distance. Therefore, the alternative which has the furthest distance to the negative-ideal solution is the most desirable. However, in the case where the Euclidean distance of two alternatives far from each other, the Taxicab distance is used to compare them.

The steps of the CODAS method are as follow:

Step 1. The step one involves the Decision-Making Matrix ($X$):

$$X = \left[x_{ij}\right]_{n \times m} = \begin{bmatrix} x_{11} & x_{12} & \cdots & x_{1m} \\ x_{21} & x_{22} & \cdots & x_{2m} \\ \cdots & \cdots & \cdots & \cdots \\ x_{n1} & x_{n2} & \cdots & x_{nm} \end{bmatrix} \tag{1}$$

where $x_{ij}$ ($x_{ij} \geq 0$) denotes the performance value of $i^{th}$ alternative on $j^{th}$ criterion. ($i \in \{1,2,\ldots,n\}$ and $j \in \{1,2,\ldots,m\}$).

Step 2. The step two requires to compute the Normalized Decision Matrix $N$. The normalized performance values $n_{ij}$ is determined as follows:

$$n_{ij} = \begin{cases} \dfrac{x_{ij}}{\max\limits_i x_{ij}} & \text{if } j \in N_b \\[3mm] \dfrac{\min\limits_i x_{ij}}{x_{ij}} & \text{if } j \in N_c \end{cases} \tag{2}$$

where $N_b$ and $N_c$ represent the sets of benefit and cost criteria, respectively

Step 3. The step three needs the Weighted Normalized Decision Matrix $R$. The weighted normalized performance values are calculated as follows:

$$R = \left[r_{ij}\right]_{n \times m} \tag{3}$$

$$r_{ij} = w_j n_{ij} \tag{4}$$

where $w_j$ ($0 \leq w_j \leq 1$) denotes the weight of $j^{th}$ criterion, and $\sum_{j=1}^{m} w_j = 1$

Step 4. The step four consists to define the Negative-Ideal Solution $NIS_j$ as follows:

$$NIS = \left[NIS_j\right]_{1 \times m} \tag{5}$$

$$NIS_j = \min_i r_{ij} \tag{6}$$

Step 5. The step five is based on the Euclidean and Taxicab distances of alternatives from the $NIS_j$ shown as follows:

$$E_i = \sqrt{\sum_{j=1}^{m} \left(r_{ij} - SIN_j\right)^2} \tag{7}$$

$$T_i = \sum_{j=1}^{m} \left|r_{ij} - SIN_j\right| \tag{8}$$

Step 6. The step six depends on structuring the Relative Evaluation Matrix $Re$, shown as,

$$Re = [h_{ik}]_{n \times n} \tag{9}$$

$$h_{ik} = (E_i - E_k) + (\psi(E_i - E_k) \times (T_i - T_k)) \tag{10}$$

where $i,k \in \{1,2,\ldots,n\}$ and $\psi$ denotes a threshold function to recognize the equality of the Euclidean distances of two alternatives, and it is defined as follows:

$$\psi(E_i - E_k) = \begin{cases} 1 \text{ if } |E_i - E_k| \geq \tau \\ 0 \text{ if } |E_i - E_k| < \tau \end{cases} \tag{11}$$

Where the threshold parameter $\tau$ is defined by the decision maker. It is suggested to set this parameter at a value between 0.01 and 0.05.

Step 7. The step seven consists to compute the Evaluation Score $ES$ of each alternative, that can be presented as follows:

$$ES_i = \sum_{k=1}^{n} h_{ik} \tag{12}$$

Step 8. The last step of the CODAS method is to rank the alternatives depending on the decreasing values of evaluation score $ES_i$.

In the evaluation stage, the alternative with the greater distance is more desirable than the others.

In the literature, only few studies are based on CODAS method in MCGDA problems. However, some extensions of CODAS method are developed. Indeed, Keshavarz Ghorabaee et al. (2017) studied the group decision-making problems within the CODAS method, which is combined with trapezoidal fuzzy numbers for market segment evaluation. Furthermore, Yeni and Ozcelik (2019) developed the interval-valued Atanassov intuitionistic fuzzy CODAS (IVAIF-CODAS) method to solve a personnel selection problem. Cherif and Frikha (2021) extended CODAS approach under uncertain environment for interval rough numbers. In the framework of CODAS parameter elicitation, Cherif and Frikha (2022a) devoleped a mathematical programming determining CODAS criteria weights. Like manner, Cherif and Frikha (2022b) determined the criteria weight values of Interval Rough CODAS approach.

## 4 The Process of an Interactive Group Preference Disaggregation Approach

For group decision process, we intend to reduce the subjectivity of the input data. Howbeit, the subjective elicitation of the threshold parameter is untreated for the CODAS method. Well, we propose a new mathematical programming model for group preference disaggregation to facilitate the determination of CODAS threshold parameter.

The incorporation of DM's preference information into MCDA approaches has been studied. The DM preferences are incorporated prior, during or after to the search process. Thus, MCDA approaches can be divided into a priori, interactively and a posteriori approach (Purshouse et al., 2014) This paper suggests an interactive disaggregation approach transacting with the DM preferences.

The CODAS approach uses the Euclidean distance as the primary measure of assessment. Besides, if two alternatives are the same Euclidean distances, then, the Taxicab distance is used to contrast them. As noted previously, the threshold parameter is too subjective since it is set by the decision maker who do not have a clear knowledge about this threshold. Also, this parameter marks the degree of closeness of Euclidean distances. So, the determination of this parameter becomes necessary to remove this uncertainty issue.

Also, CODAS is an outranking distance-based method, which use outranking relations between pairs of alternatives. Thus, the key for the disaggregation approach is to determine the CODAS threshold parameter from the binary outranking relations given by DMs. Besides, we propose an elicitation model based on Mixed Integer Linear Programming (MILP), which can now be written as follows:

*Program*

$$\text{Maximize } Z = \sum_{i=1}^{p} D_i \tag{1}$$

Subject to Constraints

$$\sum_{k=1}^{n} [(E_a - E_k) + \psi(a,\ k) * (T_a - T_k)] - \sum_{k=1}^{n} [(E_b - E_k) + \psi(b,\ k)$$

$$*(T_b - T_k)] - D_i = 0; \forall a,b \in A; \forall i = 1, \ldots ,p \tag{2}$$

$$\sum_{a=1}^{n} \sum_{b=1}^{n} \psi(a, b) \geq 1 \forall a \neq b; a = 1,..,n \text{ and } b = 1,..,n \tag{3}$$

$$\psi(a, b) = \psi(b, a) \forall a \neq b; a = 1,..,n \text{ and } b = 1,..,n \tag{4}$$

$$\psi(a, a) = 0 \forall a = 1,..,n \tag{5}$$

If $|E_a - E_b| \leq |E_c - E_d|$ then $\psi(a, b) \leq \psi(c, d) \forall a \neq b; a = 1,..,n \text{ and } b = 1,..,n$

$$\forall c \neq d; c = 1,..,n \text{ and } d = 1,..,n \tag{6}$$

If $|E_a - E_b| = |E_c - E_d|$ then $\psi(a, b) = \psi(c, d) \forall a \neq b; a = 1,..,n \text{ and } b = 1,..,n$

$$\forall c \neq d; c = 1,..,n \text{ and } d = 1,..,n \tag{7}$$

$$\psi(a, b) \in \{0, 1\} \forall a \neq b; a = 1,..,n \text{ and } b = 1,..,n \tag{8}$$

$$\tau \in [0.01, 0.05] \tag{9}$$

If $|E_a - E_b| < 0.01$ then $\psi(a, b) = 0 \forall a \neq b; a = 1,..,n \text{ and } b = 1,..,n \tag{10}$

If $|E_a - E_b| > 0.05$ then $\psi(a, b) = 1 \forall a \neq b; a = 1,..,n \text{ and } b = 1,..n \tag{11}$

$$\psi(a, b) = \begin{cases} 1 \text{ if } |E_a - E_b| \geq \tau \forall a \neq b; a = 1, .., n \text{ and } b = 1, .., n \\ 0 \text{ if } |E_a - E_b| < \tau \forall a \neq b; a = 1, .., n \text{ and } b = 1, .., n \end{cases} \tag{12}$$

$$\tau > 0 \tag{13}$$

$$D_i \geq \varepsilon_i; \forall i = 1,..,p \tag{14}$$

Where

- $\psi_{(a, b)}$ is threshold function to know the equality of Euclidean distances of two alternatives
- $E_a$ is the Euclidean distance of the alternative a
- $T_a$ is the Taxicab distance of alternative a
- $p$ is the number of the DM's preferences relations.
- $n$ is the number of alternatives.
- $\varepsilon_i$ is a minimum threshold fixed for each $D_i$.

The description of the above program is given as follows:

In the CODAS method, all the alternatives are sorted in descending order of Evaluation Scores ES. Firstly, the DM choose the pairwise preferences relations of the alternatives. Suppose the partial information defined by the DM is: "alternative a is preferred to alternative b" can be formed as $a \succ b$ if and only if $ES(a) > ES(b)$ thus $ES(a)–ES(b) > 0$. The degree of preference $D_i$ of action a compared to action b is the gap between two Evaluation Scores ES of two alternatives a and b against all criteria, that is $ES(a)–ES(b) = D_i$ for each preference i given by the DM.

To satisfy the DM's preferences and to avoid indifference relations between two alternatives, we estimate that the preference relations provided by the DM are the goals to realize. At that time, we must maximize the sum $D_i$ fixed in the objective function (1).

Furthermore, within CODAS method, as we said before, action $a$ is preferred to action $b$ $(a \succ b)$ is equivalent to Evaluation Scores $ES$ of $a$ is greater than that of $b$ $(ES(a) > ES(b))$ this signifies that $ES(a)–ES(b) > 0$. The inequality constraint $ES$ $(a)–ES(b) > 0$ can be converted to an equality constraint $ES(a)–ES(b)–D_i = 0$ and we introduce in Mixed Integer Linear Program (MILP) the additional variable, (slack variable) $D_i$ which is a positive or zero variable (non-negative variable in MILP). Also, for each constraint $i$ of superior type, we subtract the variable $D_i$, which subjected to the constraint $D_i \geq 0$.

To be two alternatives different and to respect $ES(a) > ES(b)$, $D_i$ must be different to 0. Hence, we ought to set a small positive threshold $\varepsilon_i$ to each $D_i$.

$ES(a)–ES(b)–D_i = 0$ with $ES(a) = \sum_{k=1}^{n} h_{ak}$

Then $\sum_{k=1}^{n} h_{ak} - \sum_{k=1}^{n} h_{bk} - D_i = 0$

Since $h_{ak} = (E_a - E_k) + (\psi(E_a - E_k) * (T_a - T_k))$

Yet; DM's strict preferences relations $(a \succ b)$ are modeled in the mathematical program as Eq. (2), $\forall a, b \in A$ and $\forall i = 1...p$

$$\sum_{k=1}^{n} (E_a - E_k) + (\psi(E_a - E_k) * (T_a - T_k)) - \sum_{k=1}^{n} (E_b - E_k)$$
$$+ (\psi(E_b - E_k) * (T_b - T_k)) - D_i = 0$$

Each modification of the information of DM's preference relations will lead to a change in the values of the decision variables of the linear program. These modifications at the threshold function $\psi_{(a,b)}$ values can lead to change the threshold value $\tau$ of the CODAS method.

Let $M$ be the Euclidean Difference Matrix, it must be different from the Relative Evaluation Matrix $Re$. For this, we must avoid the risk of all $\psi_{(a,b)}$ values in the matrix being zero. Hence, we have inserted the constraint (3).

Besides, the Euclidean Difference Matrix $M$ is an antisymmetric square matrix $(n \times n)$ with zero diagonal whose transpose is $-M$. Moreover, the threshold function matrix $\psi$ is formed based on the absolute value of the Euclidean differences $(|E_a E_b|)$.

So, the threshold function matrix $\psi$ will be symmetrical compared to the principal diagonal. This means that $\psi(a,b)$ and $\psi(b,a)$ take the same value as indicated in the constraint (4).

Moreover, the diagonal of the matrix $M$, $M_{(a,a)} = E_a–E_a$ is equal to zero. Therefore, the diagonal of the matrix $\psi(a, a)$ is equal to zero as it shows the constraint (5).

The comparison of threshold function concerning the differences of Euclidean distances is illustrated in constraints (6) and (7).

To simplify, we denote by $D\_E(a,b)$ the absolute value of the difference between two Euclidean distances for alternatives $a$ and $b$, which is equivalent to $D\_E(a,b) = |E_a–E_b|$.

**Fig. 1** The threshold
function representation

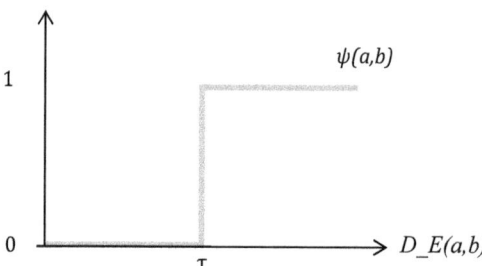

The model illustrates two mutually exclusive constraints represented in the sixth step in CODAS method: the first or the second of the two constraints must be satisfied, but not both keep simultaneously, denote either $\psi(a,b) = 0$; or $\psi(a,b) = 1$ as illustrated in Fig. 1.

According to the threshold function representation $\psi$, there are three possible cases:

*The First Case*:

If $D\_E(a,b) \geq \tau$ then $\psi(a,b) = 1$ and we suppose that $D\_E(c,d) \geq D\_E(a,b)$ then $\psi(c,d) = 1$. Therefore $\psi(a,b) = \psi(c,d) = 1$.

*The Second Case*:

If $D\_E(a,b) < \tau$ then $\psi(a,b) = 0$ and we assume $D\_E(c,d) < D\_E(a,b)$ then $\psi(c,d) = 0$. Thus $\psi(a,b) = \psi(c,d) = 0$.

*The Third Case*:

If $D\_E(a,b) < \tau \leq D\_E(c,d)$ where $\psi(a,b) = 0$ and $\psi(c,d) = 1$ then $\psi(a,b) < \psi(c,d)$.

Therefore, through these three cases, we deduce that $\psi(a,b) \leq \psi(c,d)$ if $D\_E(a,b) \leq D\_E(c,d)$ (constraint (6)) and that two pairs of alternatives having the same Euclidean difference will necessarily have the same value of the threshold functions, that is $\psi(a,b) = \psi(c,d)$ if $D\_E(a,b) = D\_E(c,d)$ (constraint(7)).

Furthermore, the CODAS method indicates that the threshold function $\psi$ is a binary variable (constraint (8)) and that threshold value $\tau$ must be between 0.01 and 0.05. For this, we orient the constraint (9).

As a coherent program, the constraint (12) and the constraint (9) help us to introduce the constraints (10) and (11).

Since $\tau$ is a decision variable, therefore non-negativity constraint is necessary for a linear program (constraint (13)).

To avoid the situation of indifference and to ensure the DM's preference information, we must define a minimum threshold $\varepsilon$ for each $D_i$ according to each strict outranking relation. Thus, we impose the non-negativity constraint that each slack variables $D_i$ is greater than a fixed threshold $\varepsilon_i$ (constraint (14)). Since Linear Programming $\varepsilon_i$ positive.

Our program illustrates logical constraint (two mutually exclusive constraints) (12) not linear constraint that need to be modeled for the linear programming model.

$$\psi(a,b) = \begin{cases} 1 \text{ if } D\_E(a,b) \geq \tau \forall a \neq b; a = 1,..,n \text{ and } b = 1,..,n \\ 0 \text{ if } D\_E(a,b) < \tau \forall a \neq b; a = 1,..,n \text{ and } b = 1,..,n \end{cases}$$

Generally, we study the logical constraints. Suppose that, in Linear Programming problem, some variables are constrained by:

$$\textbf{If } f(x_1, x_2, \ldots, x_n) > 0 \textbf{ then } g(x_1, x_2, \ldots, x_n) \geq 0 \tag{17}$$

This constraint (17) is equivalent to:

$$\begin{cases} -g(x_1, x_2, \ldots x_n) \leq My \\ f(x_1, x_2, \ldots x_n) \leq M(1-y) \\ y \in \{0, 1\} \\ M \geq Sup\{f(x_1, x_2, \ldots x_n), -g(x_1, x_2, \ldots x_n) \end{cases} \tag{18}$$

With $y$ is a binary variable (0 or 1) and M is a very large positive number verifing $M \geq f$ and $M \geq g$

We have if $\tau > D\_E(a,b)$ then $\psi(a,b) = 0$ which means that

$$\text{if } \tau - D\_E(a, b) > 0 \text{ then } \psi(a, b) \leq 0 \tag{19}$$

Hence,

$$\text{if } \tau - D\_E(a, b) > 0 \text{ then } -\psi(a, b) \geq 0 \tag{20}$$

with $\tau - D\_E(a, b) = f$ and $-\psi(a, b) = g$.
The previous system (18) is equivalent to:

$$\begin{cases} \psi(a, b) \leq My \\ \tau - D\_E(a, b) \leq M(1-y) \\ y \in \{0, 1\} \\ M \geq sup(\tau - D\_E(a, b), \psi(a, b)) \end{cases} \quad \forall a \neq b; a = 1, .., n \text{ and } b = 1, .., n \tag{21}$$

The meaning of the binary variable is there:

$$\begin{cases} y = 1, \text{ if the first constraint is satisfied and the second is rejected;} \\ y = 0, \text{ if the second constraint is satisfied and the first is rejected.} \end{cases}$$

$M$ is a very large number verify: $M \geq \tau - D\_E(a, b)$ and $M \geq -\psi(a, b)$.

Since $\tau - D\_E(a,b)$ and $-\psi(a,b)$ are less than 1, we can choose $M = 1$. Therefore, our constraint is written as follows:

$$\begin{cases} \psi(a,\ b) \leq y \\ \tau - D\_E(a,\ b) \leq 1 - y \ \forall a \neq b; a = 1,..,n \text{ and } b = 1,..,n \\ y \in \{0,\ 1\} \end{cases} \quad (22)$$

The values of $\psi(a,\ b)$ obtained from mathematical program and constraint (22) lead us to determine the value of the threshold $\tau$.

## 5 Resolution of the Mathematical Program

After modeling the model presented above, we pass to the resolution phase. It exists very efficient tools to solve linear programs, especially by IBM ILOG CPLEX software, which generates objective threshold parameter.

Since the threshold parameter elicitation model is a linear program, it is obvious to use a computer optimization tool, the modeling language OPL (Optimization Programming Language), for solving the mathematical program.

The goal is to assign the values to the decision variables so that all the constraints are satisfied (consistent). However, the preference information provided by each DM is inconsistent with the constraints of the preference disaggregation model, where the findings obtained are inconsistent. Hence, this interactive disaggregation approach was chosen to reduce the cognitive effort of the decision makers and also to cancel out the inconsistencies in their assignments. In case of inconsistency, contradictory results are presented to each DM to help him to modify his previous information.

To determine a good threshold parameter value for CODAS method, the disaggregation process needs to be repeated one or more times. In reality, the DM preferential information is not stable, he can modify it at any time, either to add or delete one or more preference relations. For this, the iterative program can model the modified information.

Once that the mathematical program is consistent, its resolution gives us the $\psi(a, b)$, $D_i$ and $\tau$. After inferring the preference threshold parameter, we apply the CODAS method to obtain the total ranking of the alternatives. Then, we present it to the decision maker. If the DM is not satisfied with the result, he can modify his information which may change over time or which may even contain contradictory and inconsistent information from the beginning. Here, it appears the role of our iterative interactive disaggregation model to assist each DM to deduce its preferences and represent them in the CODAS method, to set the new values that will be submitted again to each decision maker.

## 6 Case Study, Result, and Discussion

This section integrates Operational Research and Water Management to check the performance of the disaggregation tool in rainwater problems in Tunisian urban areas. Really, rainwater resource control has appeared for several years as an alternative solution to water drainage through sewerage networks. Besides reducing the negative effects of urban flooding and pollution, rainwater management becomes necessary to confuse the difficulties confronted. Also, decision aid tools guide urban water managers towards the best possible solution(s). So, Multi-criteria decision analysis methods find their place in the development of such decision support processes and become very fruitful.

To start the CODAS algorithm, we have to define $n$ alternatives, $m$ criteria and the weight of criteria $W_j$ ($j = 1..m$).

For reducing the volume of overflow and treating pollutants in rainwater runoff, the group decision makers are agree for eight techniques of rainwater Management $A_i(i = 1,2...,8)$, i.e., open-air pools (in water), open-air dry pools, underground basins, the valleys, porous pavements with reservoir structure, the infiltration trenches, storage roofs, interception wells. There were eight evaluation criteria $C_i(i = 1,2,...,8)$ based on those used by Martin and Legret (2005), i.e., Pollution retention ($C_1$), Probability of dysfunction ($C_2$), Needs and frequency of maintenance operations ($C_3$), Impact on groundwater quality ($C_4$); Level of approval ($C_5$), Contribution to development policies ($C_6$), Investment cost ($C_7$), Maintenance cost ($C_8$). $C_2$, $C_3$, $C_4$ as well $C_8$ are cost criteria (where lower value is preferable). These eight criteria have been evaluated by the group decision makers $DM_p$ ($p = 1,2,3$) on the basis of the analysis of the results of a satisfaction survey on the use of alternative techniques in rainwater sanitation and they provide the decision matrix presented as follow:

$$
D = \begin{array}{c} \\ A_1 \\ A_2 \\ A_3 \\ A_4 \\ A_5 \\ A_6 \\ A_7 \\ A_8 \end{array}
\begin{array}{cccccccc}
C_1 & C_2 & C_3 & C_4 & C_5 & C_6 & C_7 & C_8 \\
\left[\begin{array}{cccccccc}
4 & 20 & 3 & 2 & 5 & 3 & 38 & 32 \\
4 & 20 & 3 & 2 & 5 & 3 & 54 & 32 \\
4 & 20 & 2 & 2 & 1 & 1 & 370 & 32 \\
4 & 40 & 3 & 2 & 3.5 & 3 & 13 & 30 \\
4 & 60 & 2 & 2 & 2.5 & 2 & 54 & 4.5 \\
4 & 60 & 2 & 2 & 2.5 & 2 & 39 & 1.2 \\
1 & 40 & 2 & 5 & 1 & 2 & 0 & 2 \\
4 & 60 & 2 & 1 & 1 & 1 & 4 & 2
\end{array}\right]
\end{array} \qquad (23)
$$

**Table 2** Normalized decision matrix N

|     | $C_1$ Max | $C_2$ Min | $C_3$ Min | $C_4$ Min | $C_5$ Max | $C_6$ Max | $C_7$ Max | $C_8$ Min |
|-----|-----|-----|-----|-----|-----|-----|-----|-----|
| $A_1$ | 1 | 1 | 0,67 | 0,5 | 1 | 1 | 0,10 | 0,04 |
| $A_2$ | 1 | 1 | 0,67 | 0,5 | 1 | 1 | 0,15 | 0,04 |
| $A_3$ | 1 | 1 | 1 | 0,5 | 0,2 | 0,33 | 1 | 0,04 |
| $A_4$ | 1 | 0,5 | 0,67 | 0,5 | 0,7 | 1 | 0,04 | 0,04 |
| $A_5$ | 1 | 0,33 | 1 | 0,5 | 0,5 | 0,67 | 0,15 | 0,27 |
| $A_6$ | 1 | 0,33 | 1 | 0,5 | 0,5 | 0,67 | 0,11 | 1 |
| $A_7$ | 0,25 | 0,5 | 1 | 0,2 | 0,2 | 0,67 | 0 | 0,6 |
| $A_8$ | 1 | 0,33 | 1 | 1 | 0,2 | 0,33 | 0,01 | 0,6 |

**Table 3** Criteria weight values

|     | $C_1$ | $C_2$ | $C_3$ | $C_4$ | $C_5$ | $C_6$ | $C_7$ | $C_8$ |
|-----|-----|-----|-----|-----|-----|-----|-----|-----|
| $DM_1$ | 0.05 | 0.12 | 0.19 | 0.05 | 0.05 | 0.12 | 0.23 | 0.19 |
| $DM_2$ | 0.15 | 0.04 | 0.04 | 0.15 | 0.21 | 0.21 | 0.1 | 0.1 |
| $DM_3$ | 0.21 | 0.08 | 0.1 | 0.21 | 0.6 | 0.16 | 0.04 | 0.04 |

**Table 4** Weighted normalized performance matrix for the first DM

|     | $C_1$ | $C_2$ | $C_3$ | $C_4$ | $C_5$ | $C_6$ | $C_7$ | $C_8$ |
|-----|-----|-----|-----|-----|-----|-----|-----|-----|
| $A_1$ | 0.05 | 0.12 | 0.13 | 0.03 | 0.05 | 0.12 | 0.02 | 0.01 |
| $A_2$ | 0.05 | 0.12 | 0.13 | 0.03 | 0.05 | 0.12 | 0.03 | 0.01 |
| $A_3$ | 0.05 | 0.12 | 0.19 | 0.03 | 0.01 | 0.04 | 0.23 | 0.01 |
| $A_4$ | 0.05 | 0.06 | 0.13 | 0.03 | 0.04 | 0.12 | 0.01 | 0.01 |
| $A_5$ | 0.05 | 0.04 | 0.19 | 0.03 | 0.03 | 0.08 | 0.03 | 0.05 |
| $A_6$ | 0.05 | 0.04 | 0.19 | 0.03 | 0.03 | 0.08 | 0.03 | 0.19 |
| $A_7$ | 0.01 | 0.06 | 0.19 | 0.01 | 0.01 | 0.08 | 0.00 | 0.11 |
| $A_8$ | 0.05 | 0.04 | 0.19 | 0.05 | 0.01 | 0.04 | 0.00 | 0.11 |

According to the decision matrix provided by the group DMs, we can have the normalized decision matrix, using formula (2). It is shown in Table 2.

Min means that the objective is to minimize the criterion $j$ and Max means that the objective is to minimize the criterion $j$.

The weights of the eight criteria were provided by each decision maker $DM_p$ ($p = 1,2,3$) as given in Table 3.

Now, these weights are used to calculate the weighted normalized performance values and the Negative-Ideal Solution $NIS_j$ of each decision maker. Thus, three Weighted Normalized Performance Matrixes are recorded in Tables 4, 5, 6. Subsequently, Euclidean and Taxicab distances of alternatives are calculated using Eqs. (7) and (8), respectively, as given in Table 7.

Each decision maker provides some binary relations (Table 8), which are integrated into mathematical program.

**Table 5** Weighted normalized performance matrix for the second DM

|       | $C_1$   | $C_2$   | $C_3$   | $C_4$   | $C_5$   | $C_6$   | $C_7$   | $C_8$   |
|-------|---------|---------|---------|---------|---------|---------|---------|---------|
| $A_1$ | 0.15    | 0.04    | 0.0268  | 0.075   | 0.21    | 0.21    | 0.01    | 0.004   |
| $A_2$ | 0.15    | 0.04    | 0.0268  | 0.075   | 0.21    | 0.21    | 0.015   | 0.004   |
| $A_3$ | 0.15    | 0.04    | 0.04    | 0.075   | 0.042   | 0.0693  | 0.1     | 0.004   |
| $A_4$ | 0.15    | 0.02    | 0.0268  | 0.075   | 0.147   | 0.21    | 0.004   | 0.004   |
| $A_5$ | 0.15    | 0.0132  | 0.04    | 0.075   | 0.105   | 0.1407  | 0.015   | 0.027   |
| $A_6$ | 0.15    | 0.0132  | 0.04    | 0.075   | 0.105   | 0.1407  | 0.011   | 0.1     |
| $A_7$ | 0.0375  | 0.02    | 0.04    | 0.03    | 0.042   | 0.1407  | 0       | 0.06    |
| $A_8$ | 0.15    | 0.0132  | 0.04    | 0.15    | 0.042   | 0.0693  | 0.001   | 0.06    |

**Table 6** Weighted normalized performance matrix for the third DM

|       | $C_1$   | $C_2$   | $C_3$   | $C_4$   | $C_5$   | $C_6$   | $C_7$    | $C_8$   |
|-------|---------|---------|---------|---------|---------|---------|----------|---------|
| $A_1$ | 0.21    | 0.08    | 0.067   | 0.105   | 0.16    | 0.16    | 0.004    | 0.0016  |
| $A_2$ | 0.21    | 0.08    | 0.067   | 0.105   | 0.16    | 0.16    | 0.006    | 0.0016  |
| $A_3$ | 0.21    | 0.08    | 0.1     | 0.105   | 0.032   | 0.0528  | 0.04     | 0.0016  |
| $A_4$ | 0.21    | 0.04    | 0.067   | 0.105   | 0.112   | 0.16    | 0.0016   | 0.0016  |
| $A_5$ | 0.21    | 0.0264  | 0.1     | 0.105   | 0.08    | 0.1072  | 0.006    | 0.0108  |
| $A_6$ | 0.21    | 0.0264  | 0.1     | 0.105   | 0.08    | 0.1072  | 0.0044   | 0.04    |
| $A_7$ | 0.0525  | 0.04    | 0.1     | 0.042   | 0.032   | 0.1072  | 0        | 0.024   |
| $A_8$ | 0.21    | 0.0264  | 0.1     | 0.21    | 0.032   | 0.0528  | 0.0004   | 0.024   |

**Table 7** Euclidian and Taxicab distances of alternatives for each DM

|       | $DM_1$  |         | $DM_2$  |         | $DM_3$  |         |
|-------|---------|---------|---------|---------|---------|---------|
|       | $E_i$   | $T_i$   | $E_i$   | $T_i$   | $E_i$   | $T_i$   |
| $A_1$ | 0.129   | 0.276   | 0.252   | 0.503   | 0.244   | 0.513   |
| $A_2$ | 0.132   | 0.288   | 0.252   | 0.508   | 0.244   | 0.515   |
| $A_3$ | 0.255   | 0.426   | 0.160   | 0.298   | 0.185   | 0.347   |
| $A_4$ | 0.096   | 0.188   | 0.213   | 0.414   | 0.216   | 0.423   |
| $A_5$ | 0.103   | 0.249   | 0.157   | 0.343   | 0.188   | 0.371   |
| $A_6$ | 0.203   | 0.379   | 0.182   | 0.412   | 0.191   | 0.399   |
| $A_7$ | 0.132   | 0.230   | 0.092   | 0.147   | 0.069   | 0.123   |
| $A_8$ | 0.135   | 0.249   | 0.174   | 0.303   | 0.234   | 0.381   |

Indeed, the resolution with CPLEX software provides the individual threshold parameter $\tau_p$ of the DM $p$ as follows (Table 9).

According to Table 9 and Eq. (12), we computed the Relative Evaluation Matrix and the Evaluation Score $ES_i^p$ ($p = 1,2,3$) of each decision maker separately. Then, we implement the group decision-making side. The individual Evaluation Score can be fused into the collective Evaluation Score $ES_i$ (Table 10). So, applying the CODAS method with the inferring threshold found help us to obtain the following

**Table 8** Binary preferences relation for each DM

| DM$_1$ | DM$_2$ | DM$_3$ |
|---|---|---|
| $A_3 \succ A_6$ | $A_4 \succ A_8$ | $A_4 \succ A_3$ |
| $A_6 \succ A_1$ | $A_6 \succ A_3$ | $A_1 \succ A_6$ |
| $A_2 \succ A_5$ | $A_1 \succ A_8$ | $A_8 \succ A_5$ |
| $A_2 \succ A_1$ | $A_5 \succ A_7$ | $A_8 \succ A_6$ |
| $A_1 \succ A_4$ | $A_2 \succ A_3$ | $A_2 \succ A_1$ |
| $A_8 \succ A_7$ | $A_2 \succ A_4$ | $A_6 \succ A_7$ |
| $A_6 \succ A_4$ | $A_1 \succ A_6$ | $A_3 \succ A_7$ |

**Table 9** The threshold values of each decision maker

| Decision Maker | Threshold parameter $\tau_p$ |
|---|---|
| DM$_1$ | 0.029 |
| DM$_2$ | 0.017 |
| DM$_3$ | 0.028 |

**Table 10** Individual and collective evaluations score matrix

| Alternatives | $ES_i^1$ | $ES_i^2$ | $ES_i^3$ | $ES_i$ | Rank |
|---|---|---|---|---|---|
| $A_1$ | −0.314 | 1634 | 1284 | 0.868 | 2 |
| $A_2$ | −0.221 | 1666 | 1294 | 0.913 | 1 |
| $A_3$ | 1798 | −0.706 | −0.310 | 0.261 | 4 |
| $A_4$ | −1201 | 0.497 | 0.405 | −0.100 | 5 |
| $A_5$ | −0.688 | −0.455 | −0.170 | −0.438 | 7 |
| $A_6$ | 1188 | 0.235 | 0.021 | 0.481 | 3 |
| $A_7$ | −0.470 | −2496 | −3107 | −2024 | 8 |
| $A_8$ | −0.330 | −0.601 | 0.583 | −0.116 | 6 |

rank of the rainwater Management techniques: $A_2 \succ A_1 \succ A_6 \succ A_3 \succ A_4 \succ A_8 \succ A_5 \succ A_7$ with which the group decision makers are satisfied.

# 7 Conclusions and Future Work

This chapter presents an operational research technique, which appears in the application of CPLEX software and mathematical programming model to the study and analysis of water resource problems. We presented an iterative and interactive disaggregation approach to deal with preferences expressed by group DMs to rank alternatives within CODAS method. However, we start with a binary preference relation of alternatives provided by each DM, then the threshold parameter elicitation model for the CODAS method is built with program to reduce the subjectivity problem. Besides, the CODAS disaggregation approach optimizes the objective function of the threshold elicitation model, implying that inferring a preference model as consistent as possible with the DM' judgments in order to determine the threshold parameter of the CODAS method due to its specific properties such as subjectivity and sensitivity. To enhance a preference disaggregation process, a Tunisian case study has been used to illustrate the preference disaggregation procedure in urban rainwater management. Eight alternatives and eight criteria are considered in this study to help group DMs to choose the best technique of rainwater management.

The proposed approach can be expanded through the combination of another subjective weighting approach to infer not only threshold's CODAS from preference relations provided by group decision makers but also elicit weights and threshold parameters. The application of preferences disaggregation approach can be also used in the uncertain decision problems, that is why we wish to develop an extension of CODAS method in the group decision context to cope with uncertainty. Besides, sensitivity analysis of CODAS' parameters would be worth exploring further for more over DM preferences.

# References

Alamanos, A., Mylopoulos, N., Loukas, A., & Gaitanaros, D. (2018). An integrated multicriteria analysis tool for evaluating water resource management strategies. *Water, 10*(12), 1795.

Almasri, M. N., & Kaluarachchi, J. J. (2005). Multi-criteria decision analysis for the optimal management of nitrate contamination of aquifers. *Journal of Environmental Management, 74*(4), 365–381.

Ammar, A., Riksen, M., Ouessar, M., & Ritsema, C. (2016). Identification of suitable sites for rainwater harvesting structures in arid and semi-arid regions: A review. *International Soil and Water Conservation Research, 4*(2), 108–120.

An, N. T., & Mousseau, V. (2002). Using assignment examples to infer category limits for the electre tri method. *Journal of Multi-Criteria Decision Analysis, 11*, 29–43.

Ben Amor, W. D., & Frikha, H. M. (2018). Hierarchical structuring for the olive trees irrigation problem in Tunisia. *Multiple Criteria Decision Making, 13*, 29–55.

Brans, J. P., & Vincke, P. (1985). A preference ranking organisation method: The PROMETHEE method for multiple criteria decision-making. *Management Science, 31*, 647–656.

Cailloux, O., Meyer, P., & Mousseau, V. (2012). Eliciting ELECTRE TRI category limits for a group of decision makers. *European Journal of Operational Research, 223*, 133–140.

Calizaya, A., Meixner, O., Bengtsson, L., & Berndtsson, R. (2010). Multi-criteria decision analysis (MCDA) for integrated water resources management (IWRM) in the Lake Poopo Basin, Bolivia. *Water Resources Management, 24*(10), 2267–2289.

Chakraborty, S., & Zavadskas, E. K. (2014). Applications of WASPAS method in manufacturing decision making. *Informatica, 25*, 1–20.

Cherif, M. R., & Frikha, H. M. (2021). An extension of codas method based on interval rough numbers for multi-criteria group decision making. *Multiple Criteria Decision Making, 16*.

Cherif, M. R., & Frikha, H. M. (2022a) Inferring criteria weight parameters in CODAS method. *International Journal of Multicriteria Decision Making, 8*(4).

Cherif, M. R., & Frikha, H. M. (2022b). Criteria weight determination within Interval Rough CODAS approach for Water Resources management. *Proceedings of the 2021 International Conference on decision aid sciences and applications (DASA'21)*.

Damart, S., Dias, L. C., & Mousseau, V. (2007). Supporting groups in sorting decisions: Methodology and use of a multi-criteria aggregation/disaggregation DSS. *Decision Support Systems, 43*, 1464–1475.

Devaud, J. M., Groussaud, G., & Jacquet-Lagrèze, E. (1980). *UTADIS: Une méthode de construc-tion de fonctions d'utilité additives rendant compte de jugements globaux*. European Working Group on Multicriteria Decision Aid, Bochum.

Dias, L. C., & Climaco, J. (2000). ELECTRE TRI for groups with imprecise information on parameter values. *Group Decision and Negotiation, 9*, 355–377.

Dimitrios, I. S., & Nikolaos, P. V. (2017). Multicriteria decision aid applications to support risk decisions in the marine environment: Locating suitable transshipment areas. *Journal of Risk Analysis and Crisis Response, 7*, 3–12.

Elleuch, M. A., Elleuch, L., & Frikha, A. (2019). A hybrid approach for water resources manage-ment in Tunisia. *International Journal of Water, 13*(1), 80–99.

Fernandes, L., Ridgley, M. A., & Hof, T. v.'t. (1999). Multiple criteria analysis integrates eco-nomic, ecological and social objectives for coral reef managers. *Coral Reefs, 18*(4), 393–402.

Frikha, H., Chabchoub, H., & Martel, J. M. (2010). Inferring criteria's relative importance coeffi-cients in PROMETHEE II. *International Journal of Operational Research, 7*, 257–275.

Frikha, H., Chabchoub, H., & Martel, J.-M. (2011). An interactive disaggregation approach inferring the indifference and the preference thresholds of PROMETHEE II. *International Journal of Multicriteria Decision Making, 1*, 365–393.

Frikha, H., Chabchoub, H., & Martel, J. M. (2017). Location of a new banking agency in Sfax: A multi-criteria approach. *International Journal of Information and Decision Sciences, 9*, 45–76.

Frikha, H., & Charfi, S. (2018). Inferring an ELECTRE I model from binary outranking relations. *International Journal of Multicriteria Decision Making, 7*, 263–275.

Ghram, M., & Frikha, H. (2018). A new procedure of criteria weight determination within the ARAS method. *Multiple Criteria Decision Making, 13*, 56–73.

Greco, S., Mousseau, V., & Słowinski, R. (2008). Ordinal regression revisited: Multiple criteria ranking using a set of additive value functions. *European Journal of Operational Research, 191*, 415–435.

Hwang, C. L., & Yoon, K. (1981). *Multiple attribute decision making: Methods and applications*. Springer.

Jacquet-Lagreze, E., & Siskos, Y. (1982). Assessing a set of additive utility functions for multicriteria decision making: The UTA method. *European Journal of Operational Research, 10*, 151–164.

Jahanshahloo, G. R., Zohrehbandian, M., Alinezhad, A., Naghneh, S. A., Abbasian, H., & Mavi, R. K. (2011). Finding common weights based on the DM's preference information. *Journal of the Operational Research Society, 62*, 1796–1800.

Karnib, A. (2004). An approach to elaborate priority preorders of water resources projects based on multicriteria evaluation and fuzzy sets analysis. *Water Resources Management, 18*, 13–33.

Keshavarz Ghorabaee, M., Amiri, M., Zavadskas, E. K., Hooshmand, R., & Antuchevičienė, J. (2017). Fuzzy extension of the CODAS method for multi-criteria market segment evaluation. *Journal of Business Economics and Management, 18*, 1–19.

Keshavarz Ghorabaee, M., Zavadskas, E. K., Olfat, L., & Turskis, Z. (2015). Multi-criteria inventory classification using a new method of evaluation based on distance from average solution (EDAS). *Informatica, 26*, 435–451.

Keshavarz Ghorabaee, M., Zavadskas, E. K., Turskis, Z., & Antucheviciene, J. (2016). A new combinative distance-based assessment (CODAS) method for multi-criteria decision-making. *Economic Computation & Economic Cybernetics Studies & Research, 50*, 25–44.

Kiss, L., Martel, J. M., & Nadeau, R. (1994). ELECCALC-an interactive software for modelling the decision maker's preferences. *Decision Support Systems, 12*, 757–777.

Kumar, A., Sah, B., Singh, A. R., Deng, Y., He, X., Kumar, P., & Bansal, R. C. (2017). A review of multi criteria decision making (MCDM) towards sustainable renewable energy development. *Renewable and Sustainable Energy Reviews, 69*, 596–609.

Labreuche, C., Mayag, B., & Duqueroie, B. (2015). Extension of the MACBETH approach to elicit an ordered weighted average operator. *EURO Journal on Decision Processes, 3*, 65–105.

Li, C. W., & Tzeng, G. H. (2009). Identification of a threshold value for the DEMATEL method using the maximum mean de-entropy algorithm to find critical services provided by a semiconductor intellectual property mall. *Expert Systems with Applications, 36*, 9891–9898.

Martin, C., & Legret, M. (2005). La Méthode Multicritère ELECTRE III: Définitions, Principe et Exemple d'Application à la Gestion des Eaux Pluviales en Milieu Urbain. *Bulletin des Laboratoires des Ponts et Chaussées*, 258–259.

Mousseau, V., Figueira, J., & Naux, J. P. (2001). Using assignment examples to infer weights for ELECTRE TRI method: Some experimental results. *European Journal of Operational Research, 130*, 263–275.

Mousseau, V., & Slowinski, R. (1998). Inferring an ELECTRE TRI model from assignment examples. *Journal of Global Optimization, 12*, 157–174.

Opricovic, S., & Tzeng, G. H. (2004). Compromise solution by MCDM methods: A comparative analysis of VIKOR and TOPSIS. *European Journal of Operational Research, 156*, 445–455.

Ozerol, G., & Karasakal, E. (2007). Interactive outranking approaches for multicriteria decision-making problems with imprecise information. *Journal of Operational Research Society, 59*, 1253–1268.

Purshouse, R. C., Deb, K., Mansor, M. M., Mostaghim, S., & Wang, R. (2014). A review of hybrid evolutionary multiple criteria decision making methods. In *Conference 2014 IEEE congress on evolutionary computation (CEC)*, 1147–1154

Raju, S. K., Duckstien, L., & Arondel, C. (2000). Multicreterion analysis for sustainable water resources planning: A case study in Spain. *Water Resources Management, 14*, 435–456.

Roy, B. (1991). The outranking approach and the foundations of ELECTRE methods. *Theory and Decision, 31*, 49–73.

Shen, J., Lu, H., Zhang, Y., Song, X., & He, L. (2016). Vulnerability assessment of urban ecosystems driven by water resources, human health and atmospheric environment. *Journal of Hydrology, 536*, 457–470.

Siskos, J. (1980). Comment modéliser les préférences au moyen de fonctions d'utilité additives. *RAIRO Recherche Opérationnelle, 14*, 53–82.

Siskos, J. (1983). Analyse de systèmes de décision multicritère en univers aléatoire. *Foundations of Control Engineering, 8*, 193–212.

Toosi, A. S., Tousi, E. G., Ghassemi, S. A., Cheshomi, A., & Alaghmand, S. (2020). A multi-criteria decision analysis approach towards efficient rainwater harvesting. *Journal of Hydrology, 582*, 124501.

Yeni, F. B., & Ozcelik, G. (2019). Interval-valued atanassov intuitionistic fuzzy codas method for multi criteria group decision making problems. *Group Decision and Negotiation, 28*, 433–452.

Zardari, N. H., Ahmed, K., Shirazi, S. M., & Yusop, Z. B. (2015). *Weighting methods and their effects on multi-criteria decision making model outcomes in water resources management*. Springer.

Zavadskas, E. K., Kaklauskas, A., & Sarka, V. (1994). The new method of multicriteria complex proportional assessment of projects. *Technological and Economic Development of Economy, 1*, 131–139.

Zavadskas, E. K., Turskis, Z., Antucheviciene, J., & Zakarevicius, A. (2012). Optimization of weighted aggregated sum product assessment. *Electronics and Electrical Engineering, 122*, 3–6.

Zhaoxu, S., & Min, H. (2010). Multi-criteria decision making based on PROMETHEE method. In *Computing, control and industrial engineering, international conference on*, Wuhan, China (pp. 416–418).

# Evaluating How the Regulatory Ecosystem Promotes Entrepreneurial Activities in Africa

**Pawoumodom Matthias Takouda, Mohamed Dia, Alassane Ouattara, and Konan Vincent De Paul Kouadio**

**Abstract** In Sub-Saharan Africa, and globally, entrepreneurship is renowned as an important engine of growth and economic development. In this chapter, we use Operations Research models to study how public institutions encourage entrepreneurship. We present a case study where Principal Component Analysis (PCA) and Data Envelopment Analysis (DEA) methodologies are applied to evaluate the impact of government regulations on the creation of new businesses. We use data from the Doing Business (DB) project of the World Bank for 40 African economies during the period 2014–2018. We propose seven novel composite evaluation measures. Using correlation and consistency analysis, we validate the proposed measures by comparing them to the Start a Business scores from Doing Business. Our subsequent analyses provide relevant robust comparisons and rankings of African economies and allow the identification of best-performing countries in our sample.

## 1 Introduction

Entrepreneurship, globally, is considered as an essential factor of economic development and growth (Gedeon, 2010; Rogge & Kolyaseva, 2022). It remains one of the most popular concepts in development economics (Adusei, 2016; Sergi et al.,

P. M. Takouda · M. Dia (✉)
Research Group in Operations, Analytics and Decision Sciences (RGinOADS), School of Business Administration, Faculty of Management, Laurentian University, Sudbury, ON, Canada
e-mail: mtakouda@laurentian.ca; mdia@laurentian.ca

A. Ouattara
Cesag Recherche Lab, Département CESAG Recherche, Centre Africain d'Études Supérieures en Gestion (CESAG), Dakar, Senegal
e-mail: alassane.ouattara@cesag.edu.sn

K. V. D. P. Kouadio
Département de Sciences économiques et de gestion, Laboratoire d'économie et de gestion de l'ouest (LEGO), Brest, France
e-mail: konan-vincent.kouadio@etudiant.univ-brest.fr

© The Author(s), under exclusive license to Springer Nature Switzerland AG 2022
H. Masri (ed.), *Africa Case Studies in Operations Research*, Contributions to Management Science, https://doi.org/10.1007/978-3-031-17008-9_5

2019) because it has been identified as a key contributor of employment, innovation and sustained economic growth and development (Méndez-Picazo et al., 2012; Meyer & de Jongh, 2018). Small and medium enterprises (SME) contribute to the creation and the sustainability of productive and decent jobs regardless of the level of the economies (ILO, 2015). Yet, defining the entrepreneurship concept is not an easy task, due to the lack of unanimity on an acceptable conceptualization. Nevertheless, one possible definition is that entrepreneurship is a set of phenomena which creates and develops ventures in a specialized exchange economy (Robbins et al., 2020). Over the years, scholars have analyzed entrepreneurship through various lenses, such as digitalization, age, corruption, governance (Ayyagari et al., 2011; Ben Youssef et al., 2021; Dutta & Sobel, 2016; Fossen & Sorgner, 2019; Lee et al., 2020; Omri, 2020). Hence, it is clear that entrepreneurship is a complex and multidimensional phenomenon. It should accordingly be considered with the appropriate tools.

Entrepreneurship in Africa is not some novel phenomena. Generally speaking, entrepreneurship in Africa, particularly in Sub-Saharan African countries, is the paramount way to give succour to the local economy. The concept has attracted a lot of interest from scholars and policymakers. According to Robson et al. (2009), countries in Sub-Saharan Africa champion the development of small-and medium-sized enterprises (SMEs). But up to today in Sub-Saharan Africa, 88% of the population is either underemployed or totally unemployed (Ajide, 2020).

According to Rogge and Kolyaseva (2022), "*an essential antecedent of entrepreneurial activity is its institutional context that comprises business regulations.*" Regulations indeed impact productivity, employment growth, trade, investment, and hence entrepreneurship. As an economic agent, the entrepreneur has to cope with constraints deriving from regulations to start, operate, and grow a business (Djankov, 2016; World Bank, 2011, 2019). Hence, every country should implement economic and social policies to the benefit of SMEs. These regulations have to be efficient and should therefore consider specific characteristics of SMEs, such as access to funding, training of entrepreneurs. They also have to achieve an equilibrium between achieving the appropriate level of control and protection (including property rights, ownership rights, investor's protections, creditor's rights), and the needs to facilitate the introduction of ventures as well as the sustainability of the existing businesses (including entry barriers, tax regulations). In one word, the institutional regulations have to create a favourable ecosystem that encourages entrepreneurial potential.

It is in this context that the "Doing Business" (DB) project of The World Bank was introduced (Djankov, 2016; World Bank, 2011, 2019). It aims at raising awareness about the issue of the impact of countries business regulations on the entrepreneurial activities (Rogge & Kolyaseva, 2022). This was achieved by tracking worldwide business reform activities and annually evaluate them through an index, the Ease of Doing Business (EDB) index based on a choice of important regulatory areas that affect the lifecycle of a business. In practice, the EDB index is a composite index (Nardo et al., 2005; Foster et al., 2013; Becker et al., 2017; Greco et al., 2019) of measures assessing ten dimensions: *Starting a Business, Dealing*

*With Constructions Permits, Getting Electricity, Registering Property, Getting Credit, Protecting Minority Investors, Paying Taxes, Trading Across Borders, Enforcing Contracts, Resolving Insolvency* (Rogge & Kolyaseva, 2022; World Bank, 2011, 2019). The EDB index serves as monitoring tools, a benchmarking tool and also means to advocate for business regulations that foster a favourable business ecosystem (Rogge & Kolyaseva, 2022). Without surprise, the choices of regulatory areas that the DB project considered important and that were included in its EDB index, did not come without controversy or critics from the scholar community. Our goal here is not to delve into these important conversations. We would refer the interested reader to Rogge and Kolyaseva (2022).

Rather, our interest in this chapter is on the methodology used to construct the EDB index. The EDB index is calculated as an arithmetic mean of ten (10) scores or indicators measuring each dimension. Each score or indicator is also computed as the arithmetic mean of a certain number of sub-indicators. Each of these sub-indicators is obtained by scaling data obtained from surveys using the best and worst observations obtained. In this regard, the score of a country for a given sub-indicator can be seen as the result of a benchmarking of all the countries according solely to that sub-indicator. There are two issues with that methodology (Rogge & Kolyaseva, 2022). The first has to do with the use of arithmetic mean as the aggregation tool. This choice corresponds to assigning the same weight to each EDB dimension. Such an approach, with one size fitting all, does not allow each country to make policy choices where they would focus on certain aspects or dimensions of entrepreneurial activities and have that recognized in the EDB indexes. Policy analysts are hence less incentivized to propose policies that would be tailored for their specific business environment, as it might result in a lower score, because all dimensions have the same weight. The second issue is related to the fact that through the scaling procedures by sub-indicator of the data collected, several benchmarking based on a single criterion are performed. That approach makes the benchmarking sensible to measurement errors. In addition, benchmarking on multiple criteria would lead to a better identification of best practices.

Thankfully, the aforementioned drawbacks can be corrected with the help of several available tools. They allow the analysts to construct composite indicators, where the latent structure of the data drives the choice of the weights used for the aggregation. Further, some of the aforesaid tools also allow for benchmarking in a multicriteria setting (Nardo et al., 2005). Composite indicators have experienced a surge during the recent years, despite the lack of formal way to define them (Greco et al., 2019). They have been used in various areas of research (Nardo et al., 2005; Greco et al., 2019) including social sciences, environment and sustainability, business, finance (Melyn & Moesen, 1991; Lovell et al., 1995; Nicoletti et al., 2000; Storrie & Bjurek, 2000; Cherchye, 2001; Mahlberg & Obersteiner, 2001; Cherchye & Kuosmanen, 2004; Cherchye et al., 2004; Cámara & Tuesta, 2014; Greyling & Tregenna, 2017; Ahamed & Mallick, 2019; Anarfo et al., 2020; Takouda et al., 2020; Ouattara et al., 2021). The most commonly used data-driven approach to the construction of composite indexes are Principal Component Analysis (PCA) and Data Envelopment Analysis (DEA).

Principal Component Analysis (PCA) is a multivariate statistical technique, frequently used in the weighting of composite indexes (Nardo et al., 2005; Greco et al., 2019). It is a statistical approach that aims at reductionism: capture the largest variance possible in the original variables (standardized) with as fewer components as possible. In the context of composite indicators, PCA provides an approach where instead of selecting aggregation weights subjectively, or assume them to be all equal, the actual data guide the determination of the weights (Nicoletti et al., 2000; Cámara & Tuesta, 2014; Greyling & Tregenna, 2017; Ahamed & Mallick, 2019; Anarfo et al., 2020). Additional, potentially subjective, decisions are to be made by the decision makers: the number of principal components that are retained and whether a rotation method shall be used. Typically, rule of thumbs exists to provide guidance when these decisions are made (Nicoletti et al., 2000).

DEA is a non-parametric methodology used to measure the relative efficiency of a collection of decision-making units (DMU) (Charnes et al., 1978; Emrouznejad & Yang, 2018; Wamba et al., 2018; Chen et al., 2019). It has been used to build composite indexes by considering only one input, set equal to one, and by considering as outputs all the indicators that must be aggregated (Nardo et al., 2005; Greco et al., 2019; Takouda et al., 2020; Ouattara et al., 2021). This approach, also known as the Benefit of the Doubt (BOD), has been used to evaluate performance in various contexts, including logistics, macroeconomic policies, labour market, and social inclusion (Melyn & Moesen, 1991; Lovell et al., 1995; Storrie & Bjurek, 2000; Cherchye, 2001; Mahlberg & Obersteiner, 2001; Cherchye & Kuosmanen, 2004; Cherchye et al., 2004; Takouda et al., 2020; Ouattara et al., 2021). Compared to Principal Component Analysis, DEA does not require the existence of correlation between variables. Like PCA, it is an endogenous approach but here the weights are determined a posteriori. Also, as we aim here at assessing the performance of countries strategies or policies, DEA allows us to determine weights that are sensitive to the individual political priorities of each country. Further, it also eliminates suspicions of bias in the selection of the weight, since the weight applied to each country are the best ones, relatively to the other counties in the sample. Finally, since DEA is also a benchmarking tool, we not only obtain individual measures useful to rank the countries, but also identify a set of reference countries that each of them should emulate to improve their performance.

Despite its popularity and its wide range of applications, DEA methodologies have several limitations. To determine the weights used to build the composite score, DEA optimizes over all possible combinations of weights. This may result in optimal sets of weights that are unrealistic. This issue can be addressed by introducing additional constraints to further restrict the values that the weights can assume. The approach is called DEA with weight restrictions (Greco et al., 2019). Several options are possible: direct weight restrictions, cone ratio restrictions, assurance region restrictions and virtual inputs and output restrictions (Angulo-Meza & Lins, 2002). Another challenge of DEA is its lack of discrimination power among the units. This is particularly true for units deemed efficient since they all have the same efficiency or score equal to one (1). When DEA is used to compute composite indexes, this limitation is particularly challenging. Post Hoc DEA models, such as

DEA with weight restrictions Super-efficiency DEA and Cross-efficiency DEA models (Doyle & Green, 1994; Angulo-Meza & Lins, 2002; Greco et al., 2019; Álvarez et al., 2020), can remedy this issue. A final issue identified about the DB data was the likely presence of measurement errors. When using DEA models, one can use Bootstrapping (Simar & Wilson, 1998, 2000; Toma et al., 2017) to alleviate the impact of these errors.

Our intended research project in this chapter is to explore the potential of both PCA and DEA to tackle the first of the aforementioned issues with the EDB index: the use of arithmetic means as the aggregation tool. More specifically, in the context of Sub-Saharan African economies, we present a case study on the application of PCA and DEA approaches to evaluate the impacts of business regulations on the Start a Business (SAB) dimension of entrepreneurial activities (Djankov, 2016; World Bank, 2011, 2019). We propose seven (7) measures for such evaluation. We validate the proposed measures using correlation and consistency analysis among them, and also between them and the DB project score for the SAB dimension. To the best of our knowledge, this is an original contribution to the applications of PCA and DEA to build composite measures in the context of the Doing Business project. In fact, we are only aware of one related contribution of DEA on data from the Doing Business (DB) project from Rogge and Kolyaseva (2022). In that paper, the authors tackle the benchmarking issues of the EDB index. They use DEA, and more specifically benevolent cross-efficiency DEA models (Doyle & Green, 1994; Angulo-Meza & Lins, 2002) with respect to the DEA-based best practices countries within each of the seven (7) economic regions as identified by DB project (World Bank, 2011, 2019), to identify and analyze opportunity sets and best practices. Hence, unlike us here focusing on Sub-Saharan Africa, they consider all the geographic areas covered by the project.

To achieve our intended research project, we use data from the Doing Business (DB) project from the World Bank. Based on data availability, a sample of forty (40) Sub-Saharan countries is selected for a five-year study period (2014–2018). We build seven (7) novel composite scores to assess the impact of government regulations on the creation of new businesses. Two of the scores are based on the standard and on the Nicoletti et al. (2000) PCA approaches. The remaining five (5) derive from DEA models, specifically the Classic CCR, the CCR with weight restrictions, the super-efficiency CCR, the Bootstrap CCR and the benevolent cross-efficiency CCR. We use correlation and consistency analyses between the seven new scores and the Start a Business scores from the DB project to validate our proposed approaches. Finally, using descriptive statistics and ANOVA, we compare and rank the countries in our sample with respect to their performance.

The rest of the chapter is organized as follows. In the next section, the PCA and DEA approaches for the construction of composite indicators are presented. Our case study follows in Sect. 3 and analyzed in the subsequent section. In Sect. 5, the limitations of the current work are discussed. We conclude in Sect. 6.

## 2   Composite Indicators

A composite index (Foster et al., 2013; Nardo et al., 2005; Becker et al., 2017; Greco et al., 2019) helps to highlight the performance of the business creation processes because it includes multiple criteria or attributes which are usually aggregated to obtain a summary numerical measure that will be used to rank the different objects (Permanyer, 2011; Wang & Wang, 2021). Thus, we use composite index to portray Entrepreneurial Ecosystem in Sub-Saharan African countries in order to rank economies accordingly. In fact, composite indicators (also known as synthetic indices or performance indices) are popular tools for assessing the performance of countries/entities in various contexts, including human development, sustainability, perceived corruption, innovation, and competitiveness (Melyn & Moesen, 1991; Lovell et al., 1995; Nicoletti et al., 2000; Storrie & Bjurek, 2000; Cherchye, 2001; Mahlberg & Obersteiner, 2001; Cherchye & Kuosmanen, 2004; Cherchye et al., 2004; Cámara & Tuesta, 2014; Greyling & Tregenna, 2017; Ahamed & Mallick, 2019; Anarfo et al., 2020). Their simplicity has further strengthened the case for their adoption in several practices (Greco et al., 2019).

Despite its popularity, the use of composite index does not come without criticism. Some scholars and practitioners have deemed the tool as being statistically meaningless, for example (Greco et al., 2019). Our intent in this chapter is not to contribute to the conversation on whether composite indexes shall be used at all, or not. Rather, our aim is to use composite indexes to measure the performances of countries when it comes to their policies to ease entrepreneurial activities. We refer the interested reader to Greco et al. (2019) and the references therein regarding this question.

There are several decisions that have to be made regarding the methodological aspects when one sought to construct composite index. Those decisions include the weighting of composite indicators, the determination of the actual weights, the type of aggregation used, etc. We refer the interested reader to Greco et al. (2019) for extensive discussions regarding these aspects. We decided to use an approach where weights are determined objectively based on the latent structure of the data, rather than subjectively by the decision makers (Nardo et al., 2005). Several methods allow for such an objective determination of weights, including Correlation Analysis (CA), Regression Analysis (RA), Principal Component and Factor analyses (PCA, FA), Data Envelopment Analysis (DEA). The latter approaches, PCA and DEA, are among the most used techniques in the literature. We described them in the next subsections. We refer the interested reader to Greco et al. (2019) for the other techniques.

## 2.1  Principal Component Analysis

Principal Component analysis (PCA) is a multivariate statistical technique, frequently used in the weighting of composite indexes. It is a statistical approach that aims at reductionism (Nicoletti et al., 2000; Nardo et al., 2005; Greyling & Tregenna, 2017; Greco et al., 2019). At its core, PCA's goal is to capture the largest variance possible in the original variables (standardized) with as fewer components as possible. These components, called *principal components*, are linear combinations of the original variables. The coefficients of these linear combinations are called the *factor loadings* of these principal components. The components are ranked in such a way that the first principal component is the one that contains the most variance of the original variables, the second one contains the second most variance of the original variables, and so on. Additional, potentially subjective, decisions are to be made by the decision makers: the number of principal components that are retained and whether a rotation method shall be used (Greco et al., 2019). Typically, rule of thumbs exists to provide guidance when these decisions are made (see Nicoletti et al., 2000).

PCA has been extensively used to determine weights for indicators to build a composite index in various areas. See the recent survey from Greco et al. (2019) for a detailed list of applications. The standard PCA approach to building synthetic indexes, used in a majority of the applications of PCA, consists in using only the first principal component to build the composite index (Greco et al., 2019). But it may occur that the amount of the variance of the original variables captured by that first component alone is not appropriate.

Therefore, more principal components may need to be incorporated in the composite index. More specifically, the composite index is calculated using the method of Nicoletti et al. (2000) as follows. Each principal component is transformed into an intermediate composite index using its factor loadings. The weight applied to each original variable to compute this intermediate composite indicator is determined by dividing the square of the corresponding factor loading by the sum of the squares of all the factor loadings.

Then, if the standard approach is used, the composite index is simply the intermediate composite index of the first principal component. Otherwise, we use the first $k$ of principal components, and the composite index is the weighted average of the related $k$ intermediate composite indexes. We use as weights to the proportion of the total variance explained by a given principal component in the total variance explained by all the $k$ principal components.

As illustrated above, the PCA-based approach to determine the weights needed to construct a composite index does not require any a priori assumption or information on the weights themselves. It is a simple, but sophisticated method, and suitable for large sets of indicators and data. However, it is important to note that the approach has some limitations. It assumes for example that the indicators assume continuous data and that there exists a linear relationship among them. In other words, there has to exist a strong correlation between the indicators. In cases where these assumptions do not hold, nonlinear PCA methods shall be used (Greyling & Tregenna, 2017; Greco et al., 2019).

## 2.2   Data Envelopment Analysis

Data Envelopment Analysis (DEA) is undoubtedly one of the most prolific decision-making techniques of the last four decades. It is one of the two primary tools used in technical efficiency measurement, together with Stochastic Frontier Analysis (SFA) (Daraio & Simar, 2007; Assaf et al., 2011). DEA has been applied in diverse sectors including Agriculture, Banking, Supply Chain Management, Public Policy. A recent survey by Emroujnezad and Yang (2018) provides an up-to-date state of the art. An interested reader can also refer to recent surveys and relevant special issues from Wamba et al. (2018), Chen et al. (2019) and Greco et al. (2019).

The first DEA model was developed to measure technical efficiencies of decision-making units (DMUs) (Charnes et al., 1978). DEA models convert the multiple inputs and outputs of the DMUs into a comprehensive measure of relative efficiency for a sample of DMUs. Further, using results from the models, one can benchmark the DMUs to identify efficient ones, without making prior assumptions on the relationships between inputs and outputs. This represents significant benefits of models over parametric models used to assess efficiencies (Banker et al., 1986, 1988).

The most frequently used of these DEA models are the CCR (Charnes et al., 1978) and BCC (Banker et al., 1984) models. The former assesses overall technical efficiencies, and the latter measures managerial efficiencies. Scale efficiencies are obtained from the ratio of overall technical efficiencies to managerial efficiencies (Banker et al., 1984). We will focus here on introducing the actual DEA models that we intend to use in our study, which is the CCR model.

If one denotes by $n$ the number of DMUs, $t$ the number of outputs, $m$ the number of inputs, $x_{is}$ the value of the input $s$ for the $DMU_i$, $y_{ir}$ the value of the output $r$ for the $DMU_i$, the overall technical efficiency score, $h_i^{CCR}$, of the $DMU_i$ is the optimal value of the following linear program (LP):

$$\text{Max}\ \ h_i^{CCR} = \sum_{r=1}^{t} \mu_r y_{ir} \tag{1}$$

$$\sum_{s=1}^{m} \nu_s x_{is} = 1 \tag{2}$$

$$\sum_{r=1}^{t} \mu_r y_{jr} - \sum_{s=1}^{m} \nu_s x_{js} \leq 0, j = 1, \cdots, n \tag{3}$$

$$\mu_r, \nu_s \geq \epsilon \qquad (4)$$

The DMU being assessed here is $DMU_i$, $\mu_r$ is the relative importance of the output $r$, $\nu_s$ is the relative importance of the input $s$, and $\epsilon$ is a small positive real number.

For the CCR model (1–4), $DMU_i$ is efficient if the corresponding optimal ratio $h_i^{CCR}$ is equal to one (100%). It is inefficient when this ratio is smaller than one. In this model, the ratio can be interpreted as the proportion of the current inputs of the DMU that should yield the current outputs, if the DMU were efficient. In other words, the inputs must be reduced by $(1 - h_i^{CCR})$ with the same level of output if the DMU wants to become efficient.

DEA can be used to build composite indexes. To achieve that goal, it suffices to take as the outputs the indicators that are intended to be aggregated, with a single input always equal to one (1). This approach for building composite indicators is also referred to the Benefit of the Doubt (BoD) approach (Cherchye, 2001; Nardo et al., 2005; Greco et al., 2019; Takouda et al., 2020; Ouattara et al., 2021). Here, using DEA, one aims to optimize over all possible combination of weights to obtain the composite score the most favourable for the unit being assessed. Hence, for example, in the context where the composite scores are used to assess the performance of countries, we obtain scores that are sensitive to the political priorities of each country (Cherchye, 2001; Nardo et al., 2005). The BoD approach has been used in various areas (see Melyn & Moesen, 1991; Lovell et al., 1995; Storrie & Bjurek, 2000; Cherchye, 2001; Mahlberg & Obersteiner, 2001; Cherchye & Kuosmanen 2004; Cherchye et al., 2004; Takouda et al., 2020; Ouattara et al., 2021). We refer the interested reader to (Greco et al., 2019) for a detailed recent review on applications of the BoD approach.

Despite its popularity and its wide range of applications, DEA methodologies have several drawbacks. First, as exhibited by model (1–4) above, DEA optimizes over all possible combinations of weights. This may result in optimal sets of weights that are unrealistic. This is particularly true in a Benefit of the Doubt approach to compute composite indicators, as a result of the maximization, all the weighting capacity would be assigned to the indicator with the highest value. Specifically, a weight of one (1) is assigned to one indicator (the one with the highest value), and the weights of all the remaining indicators are set to 0. Again, in the context of composite indicators, one can encounter situations where there are multiple equilibria, multiple optimal solutions. This results in a large number of units deemed efficient, meaning they are assigned a composite score of one. Fortunately, there are simple solutions to correct these drawbacks. One of them is using additional constraints to further restrict the values that the weights can assume. The approach is called DEA with weight restrictions (Greco et al., 2019). Several options are possible: direct weight restrictions, cone ratio restrictions, assurance region restrictions and virtual inputs and output restrictions (Angulo-Meza & Lins, 2002). More precisely:

DEA with direct weights restrictions consists in adding to the model (1–4) constraints of the type:

$$\alpha_s \leq \nu_s \leq \beta_s \quad \text{for input } s$$
$$\alpha_r \leq \mu_r \leq \beta_r \quad \text{for output } r$$

where $\alpha_s$, $\alpha_r$, $\beta_s$, $\beta_r$ are selected by the analyst or provided by experts;

DEA with cone ratio weights restrictions consists in adding to the model (1–4) constraints of the type:

$$\nu_s = (A\alpha)_s \quad \text{for input } s$$
$$\mu_r = (B\beta)_r \quad \text{for output } r$$

With A (resp. B) a ($m \times k$) (resp. ($t \times l$)) matrix, and $\alpha$, $\beta$ vectors in $\mathbb{R}^k$, $\mathbb{R}^l$ respectively where $k$, $l$ are given integers;

DEA with assurance region restrictions consists in adding to the model (1–4) constraints of the type

$$\alpha_s \nu_1 \leq \nu_s \leq \beta_s \nu_1 \quad \text{for input } s$$
$$\alpha_r \mu_1 \leq \mu_r \leq \beta_r \mu_1 \quad \text{for output } r$$

where $\alpha_s$, $\alpha_r$, $\beta_s$, $\beta_r$ are provided by experts and $\nu_1$, $\mu_1$ are the weights for the first input and output respectively;

DEA with virtual weights restrictions consists in adding to the model (1–4) constraints of the type

$$\alpha_s \leq \frac{\nu_s x_{is}}{\sum_s \nu_s x_{is}} \leq \beta_s \quad \text{for input } s$$
$$\alpha_r \leq \frac{\mu_r y_{ir}}{\sum_r \mu_r y_{ir}} \leq \beta_r \quad \text{for output } r$$

where $\alpha_s$, $\alpha_r$, $\beta_s$, $\beta_r$ are selected by the analyst or provided by experts.

Note that the assurance region restrictions are a special case of the cone ratio ones, and they correspond to ensuring that the ratio between any input (respectively output) weights and the one of the first input (respectively output) stay within a given range. In our case study later, we use the assurance region restrictions.

Another challenge of DEA is its lack of discrimination power among the units. This is particularly true for units deemed efficient since they all have the same efficiency score equal to one. When DEA is used to compute composite indexes, this limitation is particularly challenging. Several advanced DEA models, called Post Hoc DEA models, can help alleviate this difficulty.

One of such Post Hoc model is the Super-efficiency DEA model (Angulo-Meza & Lins, 2002; Greco et al., 2019; Alvarez et al. 2020). It consists in a modification of the model (1–4), and especially the constraint (3). That constraint is the one that restricts efficiency score to be less than or equal to one (100%) for all DMUs. The idea here is to modify that constraint by requiring it to be satisfied for all DMUs *except the one (DMU$_i$) that is being assessed.* We obtain the model (5–8) below (Angulo-Meza & Lins, 2002; Alvarez et al. 2020).

$$\text{Max } h_i^{SE} = \sum_{r=1}^{t} \mu_r y_{ir} \tag{5}$$

$$\sum_{s=1}^{m} \nu_s x_{is} = 1 \tag{6}$$

$$\sum_{r=1}^{t} \mu_r y_{jr} - \sum_{s=1}^{m} \nu_s x_{js} \leq 0, j = 1, \cdots, n; j \neq i \tag{7}$$

$$\mu_r, \nu_s \geq \epsilon \tag{8}$$

Hence, if DMU$_i$ was not efficient previously ($h_i^{CCR} < 100\%$), its super-efficiency score obtained from (5–8) is the same as the CCR efficiency score from (1–4). If the DMU was efficient previously ($h_i^{CCR} = 100\%$), its super-efficiency score obtained from (5–8), since it is no longer restricted, can be greater than or equal to 100%. And, hence, there is no longer a lack of discrimination issues due to all efficient units all having a score of 100%.

Another approach to increase discrimination among efficient DMUs consists in computing their cross-efficiency (Doyle & Green, 1994; Angulo-Meza & Lins, 2002; Alvarez et al. 2020). The idea is the following. Through the DEA model (1–4), the unit being assessed (DMU$_i$) performs a self-evaluation against the other units $j = 1, \ldots, n, j \neq i$, and that self-evaluation is obtained using the optimal weights $\mu_r^i, \nu_s^i$. We can then define the following quantity:

$$E_{ij} = \frac{\sum_{r=1}^{t} \mu_r^i y_{jr}}{\sum_{s=1}^{m} \nu_s^i x_{js}}, i = 1, \cdots, n; j = 1, \cdots, n.$$

The number $E_{ij}$ is called the cross-efficiency of DMU$_j$ using the weighting scheme of DMU$_i$. Therefore, the number $E_{ii}$ is exactly the optimal solution $h_i^{CCR}$ from model (1–4). Recall that $\sum_{s=1}^{m} \nu_s^i x_{is} = 1$ from constraint (2) and $h_i^{CCR} = \sum_{r=1}^{t} \mu_r^i y_{ir}$.

The cross-efficiencies of DMU$_j$ ($j = 1, \cdots n$) using the weighting scheme of DMU$_i$ ($i = 1, \cdots n$) form a matrix $E = (E_{ij})_{i, j}$. From E, the cross-efficiency score of the DMU$_i$ is calculated as the average of the quantities $E_{ij}$ along the row $i$, or in other words, the average of all cross-efficiencies calculated using the optimal weighting scheme of DMU$_i$. It is also possible to calculate the average of all the quantities $E_{ij}$, excluding $E_{ii}$. Note that all the quantities $E_{ij}$ along the row $i$ are less than or equal to $E_{ii}$. Hence, the cross-efficiency score of the DMU $i$ is smaller than its self-efficiency $E_{ii}$.

In practice, the optimal weight $\mu_r^i, \nu_s^i$ obtained from model (1–4) are often not unique. As a result, we may obtain different values for the cross-efficiency score of the DMU$_i$ for the same set of the DMUs. To fix this issue, secondary goals must be added when one calculates these cross-efficiency scores of a DMU (Angulo-Meza & Lins, 2002; Alvarez et al. 2020). The most common approach is to use benevolent (respectively aggressive) approaches, which consists in maximizing (respectively minimizing) the sum of all cross-efficiencies of DMU$_j$ using the weighting scheme of DMU$_i$, subject to two constraints:

(a) The cross-efficiency of DMU$_i$ using the weighting scheme of DMU$_i$ remains equal to the optimal solution $h_i^{CCR}$ from model (1–4).
(b) No cross-efficiency of DMU$_j$ using the weighting scheme of DMU$_i$ is greater than one.

Both benevolent and aggressive approaches as described above results in a nonlinear optimization problem since its objective function is the sum of ratios functions. It can, however, be linearized into the following linear program for the Benevolent approach (Angulo-Meza & Lins, 2002; Alvarez et al. 2020).

$$\text{Max } E_i = \sum_{j \neq i} \sum_{r=1}^{t} \mu_r^i y_{jr} - \sum_{j \neq i} \sum_{s=1}^{m} \nu_s^i x_{js} \tag{9}$$

$$\sum_{s=1}^{m} \nu_s^i x_{is} = 1 \tag{10}$$

$$\sum_{r=1}^{t} \mu_r^i y_{ir} - E_{ii} \sum_{s=1}^{m} \nu_s^i x_{is} = 0 \tag{11}$$

$$\sum_{r=1}^{t} \mu_r^i y_{jr} - \sum_{s=1}^{m} \nu_s^i x_{js} \leq 0 \ \forall j \neq i, j = 1, \cdots, n \tag{12}$$

$$\mu_r^i, \nu_s^i \geq 0 \tag{13}$$

The quantity $E_i$ is the benevolent cross-efficiency score for DMU$_i$. The linear program corresponding to the aggressive approach is the minimization of the objective function in (9) subject to the constraints (10–13).

A final drawback of classic DEA models comes from the fact that they are deterministic models. Hence, they are limited by the fact that they do not consider the uncertainty (stochastic error), which is naturally present in real-life applications. From previous analysis (Toma et al., 2017), it is well known that point efficiency estimators yielded by DEA models are not consistent. A solution to this issue comes from the application of the bootstrapping methodology in frontier models.

Bootstrap methodologies are well-known and frequently used statistical resampling tools useful when one needs to perform statistical inferences for complex problems. This idea, proposed by Simar and Wilson (1998), allows the construction of confidence intervals and the generation of robust efficiency scores (Simar & Wilson, 2000). In our DEA models (1–4), we assume that the sample of inputs and outputs used, $\chi = \{(x_{js}, y_{jr}|s = 1...m; r = 1...t; j = 1...n)\}$, is a particular realization of an unknown data generating process, $P$. As a result, the true efficiency scores are also unknown. So, the optimal values, $h_i^{CCR}(i = 1, \cdots, n)$ of model (1–4) are point estimates of the true efficiency scores. Since $P$ is not available, we use the sample $\chi$ and bootstrapping to construct an approximate data generating process $\widehat{P}$ of $P$. Then, we generate $B$ samples of inputs and outputs from $\widehat{P}$, and for each sample, we solve model (1–4) to obtain B estimates $\widehat{h}_i^{b,CCR}(i = 1, \cdots, n; b = 1, \cdots, B)$. We therefore obtain a much better approximation to the actual efficiency scores as compared to the ordinary CCR DEA efficiency scores $h_i^{CCR}(i = 1, \cdots, n)$ (Simar & Wilson, 1998). In our case study below, we run 2000 iterations of this procedure to ensure enough convergence of the confidence intervals.

## 3 Case Study: Venture Initiations

Since 2003, the World Bank publishes yearly the "Doing Business" (DB) report, and the corresponding database, containing an analysis of the business climate or environment for Small and Medium Enterprises (SME) in several countries in the world (Djankov, 2016; World Bank, 2011, 2019). More specifically, the report assesses how the regulatory ecosystem promotes entrepreneurial activities. Indeed, entrepreneurs face a business climate that is more or less restrictive.

To be able to make the assessment, the DB project models entrepreneurial activities as a complex multidimensional concept (Djankov, 2016; World Bank, 2011, 2019). At the top level, entrepreneurial activity is defined through ten dimensions. An indicator for each dimension is calculated, and the DB score of an economy is the average of the ten indicators. The ten dimensions are:

1. Starting a Business
2. Dealing with Constructions permits
3. Getting Electricity
4. Registering Property
5. Getting Credit
6. Protecting Minority Investors
7. Paying Taxes
8. Trading across Borders
9. Enforcing Contracts
10. Resolving Insolvency

Each of these ten dimensions is itself further defined through a certain number of subdimensions. For example, the Starting a Business (SAB) dimension is further structured through the four subdimensions: Procedures (number) that must be completed, the Time (calendar days) needed, the Cost (percentage of the economy's income per capita) incurred and the Paid-in minimum capital (percentage of the economy's income per capita) representing the amount that the entrepreneur needs to deposit in a bank or with a third-party before registration, or in the immediate period (up to 3 months) after incorporation. On the other hands, the Getting Credit subdimensions, for example, are the legal rights of borrowers and lenders and the credit information (World Bank, 2019).

Information about each subdimension is collected using questionnaire instruments developed by The World Bank (Djankov, 2016; World Bank, 2011, 2019). For each subdimension, one of more sub-indicators are calculated from the collected data. There are a total of 41 sub-indicators for the ten indicators. For each sub-indicator, the data collected are rescaled using the following linear transformation:

$$\frac{(worst\ score - current\ score)}{(worst\ score - best\ score)},$$

which considers the best and worst scores for that subdimension for up to a five-year period. To reduce the effect of extreme outliers in the distribution of the rescaled data, the worst score is calculated after an eventual removal of outliers determined based on the dispersion of the distribution of data and only when sub-indicators were not bounded by construction or definition.

Finally, the indicators measuring each of the ten dimensions were calculated as the average of the sub-indicators. And the DB composite score is calculated by averaging the ten indicators. In other words, the DB scores and its ten indicators are calculated as composite scores using a simple average, essentially assigning the same weights to each indicator and sub-indicators. Hence, our research project which intends to explore the potential of data-driven methods to compute alternate composite scores. We expect through the use of these methods and their respective characteristics to be able to perform a better in-depth analysis of the "Doing Business" data and derive a better understanding of how the regulatory ecosystem promotes entrepreneurial activities.

We have therefore proposed the following case study as an initial step of the overall project. We will focus on only one of the ten dimensions of the entrepreneurial activities: the Starting a Business one. This corresponds to the fundamental step in entrepreneurship. Understanding how regulations regarding this dimension promote entrepreneurial activities will provide us with a great insight into the overall ecosystem. Hence, we will consider the sub-indicators:

- Score-Procedures—Men (number) [SPM];
- Score-Time—Men (days) [STM];
- Score-Cost—Men (% of income per capita) [SCM];
- Score-Procedures—Women (number) [SPW];
- Score-Time—Women (days) [STW];
- Score-Cost—Women (% of income per capita) [SCW];
- Score-Paid-in Minimum capital (% of income per capita) [SPMC]

We will focus on Sub-Saharan African economies. This choice stems from our interest as researchers to promote Operations Research-based scholarly works focusing on challenges in Africa. Developments of entrepreneurial activities and implementation of efficient regulations that promote them is central to many of the policies (national, regional, and international) that are in place to lead to the development of the continent.

Our period of study was between the years 2014–2018, the 5 years following the last reform of the Doing Business Database. Based on data availability, we obtain a sample of 40 sub-Saharan economies, listed in the first columns of Tables 2 and 3 below. More specifically, we have a sample of 200-panel data from 40 countries and 5 years.

Tables 1, 2 and 3 below provide the descriptive statistics of the seven (7) sub-indicators, as well as the Start a Business score (SSB) of the DB project. Table 1 provides the measures yearly for the sample. Table 2 shows the measures by economies, for countries which names start with letter A to L. Table 3 indicate the same measures for country which names start with letter M to Z.

It appears from Table 1 that there has been a yearly improvement for SSB and each of the seven sub-indicators. Tables 2 and 3 indicate two recurring behaviours. First, several countries exhibit constant values for some sub-indicators for the 5 year period. This indicates that there has been no improvement (nor deterioration) for those sub-indicators. Further, for several countries, the value of the sub-indicator, Paid in Minimum Capital (SPMC) has been constantly equal to 100, the maximum value, for each of the 5 years. Given that we intend to use DEA, this is indicative of probable issues: a weight of one (1) is assigned to the sub-indicator SPMC and the weights of all the remaining indicators are set to 0; as well as situations where there are multiple equilibria, multiple optimal solutions.

We propose seven (7) alternatives composites measures of the Start a Business dimension. The first two, denoted ICE1 and ICE2 are computed using a Principal Component Analysis (PCA) approach. ICE1 is based on standard PCA approach, and ICE2 uses all the principal components for its construction, following the method of Nicoletti et al. (2000). Our choice is motivated as follows. We aim at considering the situation where the number of principal components considered a priori. And we are hence looking at the two extreme possibilities. The remaining five (5) measures are based on DEA models. We use the classical CCR DEA models to build ICE3, CCR DEA with weight restrictions for ICE4, super-efficiency CCR DEA for ICE5, Bootstrap DEA for ICE6, and benevolent cross-efficiency for ICE7. The use of DEA with weight restrictions, super-efficiency DEA, and benevolent cross-efficiency DEA is justified by the observations made at the end of the previous paragraph, mainly some values of SPMC are equal to the maximum value 100. We also apply Bootstrap DEA to account for the likely existence of uncertainty in the data given that they were collected through a survey.

The Principal Component Analysis is performed using *IBM SPSS Statistics 24* (IBM, 2016). All the DEA models, except the Bootstrap DEA are solved using the software *EMS* (Scheel, 2000). The DEA with bootstrapping is solved using *R programming* (R Core Team, 2017).

**Table 1** Sub-indicators of the Starting a Business dimension—Descriptive statistics per year

|      | Year | 2014 | 2015 | 2016 | 2017 | 2018 | Sample |
|------|------|------|------|------|------|------|--------|
|      | N | 40 | 40 | 40 | 40 | 40 | 200 |
|      | Minimum | 34.28 | 41.86 | 41.92 | 46.36 | 50.26 | 34.28 |
| SSB  | Average | 69.32 | 72.66 | 75.04 | 77.11 | 78.80 | 74.59 |
|      | St. Dev. | 14.35 | 12.63 | 12.06 | 11.76 | 10.55 | 12.66 |
|      | Maximum | 93.74 | 94.25 | 94.51 | 94.45 | 92.00 | 94.51 |
|      | N | 40 | 40 | 40 | 40 | 40 | 200 |
|      | Minimum | 23.53 | 23.53 | 23.53 | 23.53 | 29.41 | 23.53 |
| SPM  | Average | 56.84 | 57.73 | 57.73 | 59.93 | 61.70 | 58.79 |
|      | St. Dev. | 16.17 | 16.36 | 16.47 | 16.17 | 15.78 | 16.13 |
|      | Maximum | 88.24 | 88.24 | 88.24 | 88.24 | 88.24 | 88.24 |
|      | N | 40 | 40 | 40 | 40 | 40 | 200 |
|      | Minimum | 2.01 | 9.05 | 9.05 | 9.05 | 16.08 | 2.01 |
| STM  | Average | 70.26 | 72.29 | 73.44 | 74.12 | 75.97 | 73.22 |
|      | St. Dev. | 24.18 | 22.32 | 21.56 | 21.81 | 19.84 | 21.73 |
|      | Maximum | 95.48 | 95.48 | 96.48 | 96.48 | 96.48 | 96.48 |
|      | N | 40 | 40 | 40 | 40 | 40 | 200 |
|      | Minimum | 6.87 | 24.71 | 24.81 | 20.10 | 14.34 | 6.87 |
| SCM  | Average | 71.71 | 75.59 | 78.12 | 80.23 | 80.76 | 77.28 |
|      | St. Dev. | 23.56 | 19.54 | 18.30 | 18.08 | 18.47 | 19.78 |
|      | Maximum | 99.86 | 99.86 | 99.87 | 99.88 | 99.88 | 6.87 |
|      | N | 40 | 40 | 40 | 40 | 40 | 200 |
|      | Minimum | 23.53 | 23.53 | 23.53 | 23.53 | 29.41 | 23.53 |
| SPW  | Average | 56.11 | 56.99 | 56.99 | 59.20 | 60.96 | 58.05 |
|      | St. Dev. | 16.14 | 16.38 | 16.49 | 16.18 | 15.82 | 16.14 |
|      | Maximum | 88.24 | 88.24 | 88.24 | 88.24 | 88.24 | 88.24 |
|      | N | 40 | 40 | 40 | 40 | 40 | 200 |
|      | Minimum | 1.01 | 9.05 | 9.05 | 9.05 | 16.08 | 1.01 |
| STW  | Average | 70.13 | 72.17 | 73.31 | 74.00 | 75.84 | 73.09 |
|      | St. Dev. | 24.21 | 22.30 | 21.55 | 21.80 | 19.18 | 21.73 |
|      | Maximum | 95.48 | 95.48 | 96.48 | 96.48 | 96.48 | 96.48 |
|      | N | 40 | 40 | 40 | 40 | 40 | 200 |
|      | Minimum | 6.87 | 24.71 | 24.81 | 20.10 | 14.34 | 6.87 |
| SCW  | Average | 71.71 | 75.58 | 78.11 | 80.22 | 80.75 | 77.27 |
|      | St. Dev. | 23.56 | 19.53 | 18.30 | 18.08 | 18.47 | 19.79 |
|      | Maximum | 99.86 | 99.86 | 99.87 | 99.88 | 99.88 | 99.88 |
|      | N | 40 | 40 | 40 | 40 | 40 | 200 |
|      | Minimum | 2.87 | 17.06 | 31.40 | 44.66 | 51.90 | 2.87 |
| SPMC | Average | 78.90 | 85.48 | 91.32 | 94.57 | 97.21 | 89.50 |
|      | St. Dev. | 30.63 | 25.68 | 18.09 | 12.83 | 8.86 | 21.64 |
|      | Maximum | 100 | 100 | 100 | 100 | 100 | 100 |

**Table 2** Sub-indicators of the Starting a Business dimension—Descriptive statistics per economies [A to L]

| | Economy | Angola | Benin | Botswana | Burkina Faso | Burundi | Cabo Verde | Cameroon | Chad | Comoros | Congo, Rep. |
|---|---|---|---|---|---|---|---|---|---|---|---|
| | N | 5 | 5 | 5 | 5 | 5 | 5 | 5 | 5 | 5 | 5 |
| SSB | Minimum | 55.63 | 54.25 | 76.20 | 69.06 | 91.94 | 81.91 | 72.60 | 34.28 | 58.24 | 48.51 |
| | Average | 69.71 | 79.36 | 76.56 | 80.24 | 93.78 | 82.02 | 75.77 | 44.05 | 66.12 | 58.38 |
| | St. Dev. | 12.20 | 14.89 | 0.78 | 10.15 | 1.07 | 0.09 | 3.84 | 7.16 | 6.76 | 5.83 |
| | Maximum | 79.67 | 90.58 | 77.95 | 88.17 | 94.51 | 82.15 | 82.39 | 51.91 | 72.01 | 63.83 |
| | N | 5 | 5 | 5 | 5 | 5 | 5 | 5 | 5 | 5 | 5 |
| SPM | Minimum | 58.82 | 64.71 | 52.94 | 88.24 | 88.24 | 52.94 | 70.59 | 52.94 | 52.94 | 41.17 |
| | Average | 60.00 | 69.41 | 52.94 | 88.24 | 88.24 | 52.94 | 70.59 | 52.94 | 52.94 | 42.35 |
| | St. Dev. | 2.63 | 6.44 | 0.00 | 0.00 | 0.00 | 0.00 | 0.00 | 0.00 | 0.00 | 2.63 |
| | Maximum | 64.71 | 76.47 | 52.94 | 88.24 | 88.24 | 52.94 | 70.59 | 52.94 | 52.94 | 47.06 |
| | N | 5 | 5 | 5 | 5 | 5 | 5 | 5 | 5 | 5 | 5 |
| STM | Minimum | 34.17 | 82.42 | 52.26 | 87.44 | 95.48 | 82.41 | 84.42 | 40.20 | 84.42 | 2.01 |
| | Average | 52.26 | 88.84 | 53.67 | 87.43 | 96.08 | 82.41 | 84.42 | 40.20 | 80.42 | 40.80 |
| | St. Dev. | 16.51 | 4.12 | 3.15 | 0.00 | 0.55 | 0.00 | 0.00 | 0.00 | 0.00 | 21.69 |
| | Maximum | 64.32 | 92.46 | 59.30 | 87.44 | 96.48 | 82.41 | 84.42 | 40.20 | 84.42 | 51.26 |
| | N | 5 | 5 | 5 | 5 | 5 | 5 | 5 | 5 | 5 | 5 |
| SCM | Minimum | 34.97 | 38.66 | 99.55 | 77.67 | 83.04 | 92.31 | 81.89 | 6.87 | 50.80 | 61.74 |
| | Average | 69.60 | 76.88 | 99.62 | 78.13 | 90.79 | 92.73 | 82.93 | 18.17 | 55.06 | 70.40 |
| | St. Dev. | 29.13 | 24.44 | 0.05 | 0.42 | 4.42 | 0.35 | 0.89 | 7.63 | 2.68 | 5.50 |
| | Maximum | 91.50 | 98.18 | 99.67 | 78.68 | 93.31 | 93.25 | 84.01 | 24.81 | 57.94 | 73.94 |

(continued)

**Table 2** (continued)

| | Economy | Angola | Benin | Botswana | Burkina Faso | Burundi | Cabo Verde | Cameroon | Chad | Comoros | Congo, Rep. |
|---|---|---|---|---|---|---|---|---|---|---|---|
| | N | 5 | 5 | 5 | 5 | 5 | 5 | 5 | 5 | 5 | 5 |
| SPW | Minimum | 58.82 | 58.82 | 52.94 | 88.24 | 88.24 | 52.94 | 64.71 | 52.94 | 52.94 | 35.29 |
| | Average | 60.00 | 63.53 | 52.94 | 88.24 | 88.24 | 52.94 | 64.71 | 52.94 | 52.94 | 36.47 |
| | St. Dev. | 2.63 | 6.44 | 0.00 | 0.00 | 0.00 | 0.00 | 0.00 | 0.00 | 0.00 | 2.63 |
| | Maximum | 64.71 | 70.59 | 52.94 | 88.24 | 88.24 | 52.94 | 64.71 | 52.94 | 52.94 | 41.18 |
| | N | 5 | 5 | 5 | 5 | 5 | 5 | 5 | 5 | 5 | 5 |
| STW | Minimum | 34.17 | 81.41 | 52.26 | 87.44 | 95.48 | 82.41 | 83.41 | 40.20 | 84.42 | 1.01 |
| | Average | 52.26 | 87.84 | 53.67 | 87.44 | 96.08 | 82.41 | 83.41 | 40.20 | 84.42 | 39.80 |
| | St. Dev. | 16.51 | 4.12 | 3.15 | 0.00 | 0.55 | 0.00 | 0.00 | 00.00 | 0.00 | 21.69 |
| | Maximum | 64.32 | 91.46 | 59.30 | 87.44 | 96.48 | 82.41 | 83.41 | 40.20 | 84.42 | 50.25 |
| | N | 5 | 5 | 5 | 5 | 5 | 5 | 5 | 5 | 5 | 5 |
| SCW | Minimum | 34.97 | 38.62 | 99.55 | 77.67 | 83.04 | 92.31 | 81.68 | 6.87 | 50.80 | 61.17 |
| | Average | 69.60 | 76.85 | 99.62 | 78.13 | 90.79 | 92.73 | 82.71 | 18.17 | 55.06 | 70.40 |
| | St. Dev. | 29.13 | 24.45 | 0.05 | 0.42 | 4.42 | 0.35 | 0.90 | 7.63 | 2.68 | 5.50 |
| | Maximum | 91.50 | 98.15 | 99.67 | 78.68 | 93.31 | 93.25 | 83.80 | 24.81 | 57.94 | 73.94 |
| | N | 5 | 5 | 5 | 5 | 5 | 5 | 5 | 5 | 5 | 5 |
| SPMC | Minimum | 94.54 | 34.69 | 100 | 22.88 | 100 | 100 | 57.05 | 37.11 | 40.74 | 76.96 |
| | Average | 96.97 | 85.77 | 100 | 67.16 | 100 | 100 | 68.70 | 64.87 | 72.05 | 83.41 |
| | St. Dev. | 2.78 | 28.56 | 0.00 | 40.22 | 0.00 | 0.00 | 15.53 | 27.06 | 27.42 | 8.98 |
| | Maximum | 100 | 98.66 | 100 | 98.34 | 100 | 100 | 95.85 | 94.40 | 92.72 | 99.27 |

| | Economy | Cote d'ivoire | Eritrea | Eswatini | Ethiopia | Gabon | Gambia | Ghana | Kenya | Lesotho | Liberia |
|---|---|---|---|---|---|---|---|---|---|---|---|
| | N | 5 | 5 | 5 | 5 | 5 | 5 | 5 | 5 | 5 | 5 |
| SSB | Minimum | 74.93 | 42.80 | 70.92 | 46.57 | 73.72 | 61.26 | 83.17 | 67.68 | 82.59 | 90.35 |
| | Average | 86.92 | 46.15 | 73.30 | 54.77 | 75.20 | 67.07 | 83.68 | 73.52 | 82.87 | 90.59 |
| | St. Dev. | 6.70 | 2.86 | 1.40 | 8.47 | 2.95 | 3.34 | 0.31 | 6.75 | 0.18 | 1.16 |
| | Maximum | 90.00 | 50.60 | 74.35 | 68.43 | 80.48 | 69.37 | 84.02 | 82.23 | 83.06 | 90.77 |
| | N | 5 | 5 | 5 | 5 | 5 | 5 | 5 | 5 | 5 | 5 |
| SPM | Minimum | 70.59 | 29.41 | 35.29 | 23.53 | 52.94 | 58.82 | 58.82 | 29.41 | 64.71 | 76.47 |
| | Average | 75.29 | 29.41 | 35.29 | 25.88 | 54.12 | 63.53 | 58.82 | 41.18 | 64.71 | 76.47 |
| | St. Dev. | 2.63 | 0.00 | 0.00 | 5.26 | 2.63 | 2.63 | 0.00 | 16.64 | 0.00 | 0.00 |
| | Maximum | 76.47 | 29.41 | 35.29 | 35.29 | 58.82 | 64.71 | 58.82 | 64.71 | 64.71 | 76.47 |
| | N | 5 | 5 | 5 | 5 | 5 | 5 | 5 | 5 | 5 | 5 |
| STM | Minimum | 92.46 | 16.08 | 62.31 | 61.31 | 53.27 | 73.37 | 86.43 | 66.33 | 71.36 | 94.47 |
| | Average | 93.07 | 16.80 | 68.74 | 64.12 | 56.08 | 74.77 | 86.43 | 72.56 | 71.36 | 94.47 |
| | St. Dev. | 0.55 | 0.00 | 3.60 | 2.70 | 6.30 | 0.90 | 0.00 | 5.34 | 0.00 | 0.00 |
| | Maximum | 93.47 | 16.08 | 70.35 | 67.34 | 67.34 | 75.38 | 86.43 | 78.39 | 71.36 | 94.47 |
| | N | 5 | 5 | 5 | 5 | 5 | 5 | 5 | 5 | 5 | 5 |
| SCM | Minimum | 77.78 | 74.74 | 86.16 | 47.51 | 92.46 | 12.86 | 88.33 | 72.96 | 94.30 | 90.45 |
| | Average | 88.16 | 80.54 | 89.26 | 59.51 | 93.79 | 29.96 | 90.09 | 80.34 | 95.40 | 91.43 |
| | St. Dev. | 5.84 | 4.23 | 2.44 | 9.40 | 1.53 | 10.05 | 1.07 | 5.89 | 0.73 | 0.63 |
| | Maximum | 91.76 | 86.51 | 91.82 | 71.08 | 96.38 | 37.42 | 91.25 | 86.84 | 96.17 | 9215 |
| | N | 5 | 5 | 5 | 5 | 5 | 5 | 5 | 5 | 5 | 5 |
| SPW | Minimum | 70.59 | 29.41 | 35.29 | 23.53 | 52.94 | 58.82 | 58.82 | 29.41 | 64.71 | 76.47 |
| | Average | 75.29 | 29.41 | 35.29 | 25.88 | 54.11 | 62.53 | 58.82 | 41.17 | 64.71 | 76.47 |
| | St. Dev. | 2.63 | 0.00 | 0.00 | 5.26 | 2.63 | 2.63 | 0.00 | 16.64 | 0.00 | 0.00 |
| | Maximum | 76.47 | 29.41 | 35.29 | 35.29 | 58.82 | 64.71 | 58.82 | 64.71 | 64.71 | 76.47 |
| | N | 5 | 5 | 5 | 5 | 5 | 5 | 5 | 5 | 5 | 5 |

(continued)

**Table 2** (continued)

| | Economy | Cote d'ivoire | Eritrea | Eswatini | Ethiopia | Gabon | Gambia | Ghana | Kenya | Lesotho | Liberia |
|---|---|---|---|---|---|---|---|---|---|---|---|
| STW | Minimum | 92.46 | 16.08 | 62.31 | 61.31 | 53.26 | 73.37 | 86.43 | 66.33 | 71.36 | 94.47 |
| | Average | 93.07 | 16.08 | 68.74 | 64.12 | 56.08 | 74.77 | 86.43 | 72.56 | 71.36 | 94.47 |
| | St. Dev. | 0.55 | 0.00 | 3.60 | 2.70 | 6.29 | 0.90 | 0.00 | 5.34 | 0.00 | 0.00 |
| | Maximum | 93.47 | 16.08 | 70.35 | 67.34 | 67.34 | 75.38 | 86.43 | 78.39 | 71.36 | 94.47 |
| | N | 5 | 5 | 5 | 5 | 5 | 5 | 5 | 5 | 5 | 5 |
| SCW | Minimum | 77.78 | 74.74 | 86.16 | 47.51 | 92.46 | 12.86 | 88.33 | 72.96 | 94.30 | 90.45 |
| | Average | 88.16 | 80.54 | 89.26 | 59.51 | 93.79 | 29.96 | 90.09 | 80.34 | 95.40 | 91.43 |
| | St. Dev. | 5.84 | 4.23 | 2.44 | 9.40 | 1.53 | 10.05 | 1.07 | 5.89 | 0.73 | 0.63 |
| | Maximum | 91.76 | 86.51 | 91.82 | 71.08 | 96.38 | 37.42 | 91.25 | 86.84 | 96.17 | 92.15 |
| | N | 5 | 5 | 5 | 5 | 5 | 5 | 5 | 5 | 5 | 5 |
| SPMC | Minimum | 58.90 | 50.98 | 99.89 | 53.95 | 95.11 | 100 | 99.08 | 100 | 100 | 100 |
| | Average | 91.14 | 58.55 | 99.92 | 69.55 | 69.82 | 100 | 99.37 | 100 | 100 | 100 |
| | St. Dev. | 18.02 | 7.31 | 0.02 | 18.04 | 1.76 | 0.00 | 0.19 | 0.00 | 0.00 | 0.00 |
| | Maximum | 99.29 | 70.38 | 99.94 | 100 | 99.36 | 100 | 99.57 | 100 | 100 | 100 |

**Table 3** Sub-indicators of the Starting a Business dimension—Descriptive statistics per economies [M to Z]

| | Economy | Madagascar | Malawi | Mali | Mauritania | Mauritius | Mozambique | Namibia | Nigeria | Rwanda | Senegal |
|---|---|---|---|---|---|---|---|---|---|---|---|
| | N | 5 | 5 | 5 | 5 | 5 | 5 | 5 | 5 | 5 | 5 |
| SSB | Minimum | 78.16 | 61.83 | 62.93 | 58.17 | 91.43 | 71.00 | 68.46 | 74.21 | 80.60 | 74.16 |
| | Average | 81.38 | 70.33 | 72.07 | 77.93 | 91.66 | 74.03 | 69.37 | 77.79 | 84.81 | 84.19 |
| | St. Dev. | 3.80 | 6.37 | 10.93 | 14.88 | 0.21 | 1.91 | 1.43 | 2.41 | 2.99 | 5.88 |
| | Maximum | 87.78 | 76.73 | 84.12 | 91.80 | 92.00 | 75.86 | 71.90 | 80.76 | 87.66 | 89.70 |
| | N | 5 | 5 | 5 | 5 | 5 | 5 | 5 | 5 | 5 | 5 |
| SPM | Minimum | 47.06 | 47.06 | 76.47 | 52.94 | 76.47 | 47.06 | 47.06 | 56.12 | 58.82 | 82.35 |
| | Average | 57.65 | 58.82 | 76.47 | 65.88 | 76.47 | 47.06 | 47.06 | 56.12 | 68.24 | 82.35 |
| | St. Dev. | 11.32 | 7.20 | 0.00 | 10.52 | 0.00 | 0.00 | 0.00 | 0.00 | 7.89 | 0.00 |
| | Maximum | 76.471 | 64.71 | 76.47 | 82.35 | 76.47 | 47.06 | 47.06 | 56.12 | 76.47 | 82.35 |
| | N | 5 | 5 | 5 | 5 | 5 | 5 | 5 | 5 | 5 | 5 |
| STM | Minimum | 87.44 | 60.30 | 86.45 | 81.41 | 94.47 | 79.40 | 34.17 | 70.05 | 93.47 | 94.47 |
| | Average | 89.05 | 62.31 | 90.45 | 90.65 | 94.67 | 80.20 | 36.58 | 73.42 | 94.87 | 94.47 |
| | St. Dev. | 1.96 | 1.23 | 1.38 | 5.24 | 0.45 | 1.80 | 5.40 | 5.08 | 1.52 | 0.00 |
| | Maximum | 92.46 | 63.32 | 91.96 | 94.47 | 95.48 | 83.42 | 46.23 | 81.49 | 96.48 | 94.47 |
| | N | 5 | 5 | 5 | 5 | 5 | 5 | 5 | 5 | 5 | 5 |
| SCM | Minimum | 76.52 | 39.97 | 60.95 | 76.93 | 98.22 | 53.53 | 92.63 | 70.67 | 70.10 | 67.83 |
| | Average | 78.84 | 60.19 | 65.46 | 87.50 | 98.95 | 68.85 | 93.82 | 81.63 | 76.13 | 71.20 |
| | St. Dev. | 2.20 | 17.70 | 4.50 | 5.92 | 0.46 | 9.27 | 0.78 | 6.17 | 5.57 | 6.66 |
| | Maximum | 8217 | 78.89 | 70.81 | 90.37 | 99.49 | 76.97 | 94.44 | 85.41 | 84.62 | 83.09 |

(continued)

**Table 3** (continued)

| | Economy | Madagascar | Malawi | Mali | Mauritania | Mauritius | Mozambique | Namibia | Nigeria | Rwanda | Senegal |
|---|---|---|---|---|---|---|---|---|---|---|---|
| | N | 5 | 5 | 5 | 5 | 5 | 5 | 5 | 5 | 5 | 5 |
| SPW | Minimum | 47.06 | 47.06 | 76.47 | 52.94 | 70.59 | 47.06 | 47.06 | 56.12 | 58.82 | 82.35 |
| | Average | 57.65 | 58.82 | 76.47 | 65.88 | 70.59 | 47.06 | 47.06 | 56.12 | 68.24 | 82.35 |
| | St. Dev. | 11.32 | 7.20 | 0.00 | 10.52 | 0.00 | 0.00 | 0.00 | 0.00 | 7.89 | 0.00 |
| | Maximum | 76.47 | 64.71 | 76.47 | 82.35 | 70.59 | 47.06 | 47.06 | 56.12 | 76.47 | 82.35 |
| | N | 5 | 5 | 5 | 5 | 5 | 5 | 5 | 5 | 5 | 5 |
| STW | Minimum | 87.44 | 60.30 | 89.45 | 81.41 | 93.47 | 79.40 | 34.17 | 70.05 | 93.47 | 94.47 |
| | Average | 89.05 | 62.31 | 90.45 | 90.65 | 93.67 | 80.20 | 36.59 | 73.42 | 94.87 | 94.47 |
| | St. Dev. | 1.96 | 1.23 | 1.38 | 5.24 | 0.45 | 1.80 | 5.39 | 5.08 | 1.52 | 0.00 |
| | Maximum | 92.46 | 63.32 | 91.96 | 94.47 | 94.47 | 83.42 | 46.23 | 81.49 | 96.48 | 94.47 |
| | N | 5 | 5 | 5 | 5 | 5 | 5 | 5 | 5 | 5 | 5 |
| SCW | Minimum | 76.52 | 39.97 | 60.95 | 76.93 | 98.22 | 53.53 | 92.63 | 70.67 | 70.10 | 67.83 |
| | Average | 78.84 | 60.19 | 65.46 | 87.50 | 98.95 | 68.85 | 93.82 | 81.63 | 76.13 | 71.20 |
| | St. Dev. | 2.20 | 17.70 | 4.50 | 5.92 | 0.46 | 9.27 | 0.78 | 6.17 | 5.57 | 6.66 |
| | Maximum | 82.17 | 78.89 | 70.81 | 90.37 | 99.48 | 76.97 | 94.44 | 85.41 | 84.62 | 83.09 |
| | N | 5 | 5 | 5 | 5 | 5 | 5 | 5 | 5 | 5 | 5 |
| SPMC | Minimum | 100 | 100 | 24.83 | 17.06 | 100 | 100 | 100 | 100 | 100 | 51.98 |
| | Average | 100 | 100 | 55.91 | 67.69 | 100 | 100 | 100 | 100 | 100 | 88.74 |
| | St. Dev. | 0.00 | 0.00 | 39.01 | 44.27 | 0.00 | 0.00 | 0.00 | 0.00 | 0.00 | 20.61 |
| | Maximum | 100 | 100 | 98.60 | 100 | 100 | 100 | 100 | 100 | 100 | 98.86 |

| | Economy | Seychelles | Sierra Leone | South Africa | Sudan | Sao Tome | Tanzania | Togo | Uganda | Zambia | Zimbabwe | sample |
|---|---|---|---|---|---|---|---|---|---|---|---|---|
| | N | 5 | 5 | 5 | 5 | 5 | 5 | 5 | 5 | 5 | 5 | 200 |
| SSB | Minimum | 78.32 | 83.68 | 79.71 | 73.17 | 69.30 | 68.47 | 47.07 | 64.19 | 84.32 | 47.92 | 34.28 |
| | Average | 78.52 | 87.01 | 80.16 | 73.83 | 73.71 | 69.93 | 73.14 | 61.75 | 84.93 | 50.86 | 74.59 |
| | St. Dev. | 0.15 | 3.71 | 0.72 | 0.77 | 2.99 | 1.70 | 14.81 | 3.05 | 0.10 | 4.74 | 12.66 |
| | Maximum | 78.68 | 91.15 | 81.43 | 75.14 | 77.33 | 72.12 | 82.51 | 71.12 | 85.09 | 59.28 | 94.51 |
| | N | 5 | 5 | 5 | 5 | 5 | 5 | 5 | 5 | 5 | 5 | 200 |
| SPM | Minimum | 52.94 | 70.59 | 64.71 | 47.06 | 64.71 | 41.18 | 64.71 | 23.53 | 64.71 | 47.06 | 23.53 |
| | Average | 52.94 | 72.94 | 65.88 | 47.06 | 69.41 | 41.18 | 71.77 | 25.88 | 64.71 | 48.24 | 58.79 |
| | St. Dev. | 0.00 | 3.22 | 2.63 | 0.00 | 2.63 | 0.00 | 4.93 | 3.22 | 0.00 | 2.63 | 16.13 |
| | Maximum | 52.94 | 76.47 | 70.59 | 47.06 | 70.59 | 41.18 | 76.48 | 29.41 | 64.71 | 52.94 | 88.24 |
| | N | 5 | 5 | 5 | 5 | 5 | 5 | 5 | 5 | 5 | 5 | 200 |
| STM | Minimum | 68.34 | 88.44 | 54.27 | 64.32 | 92.46 | 71.36 | 81.41 | 72.36 | 91.96 | 9.045 | 2.01 |
| | Average | 68.34 | 90.45 | 54.87 | 64.32 | 93.27 | 71.36 | 90.25 | 73.77 | 91.96 | 15.08 | 73.22 |
| | St. Dev. | 0.00 | 2.01 | 0.55 | 0.00 | 0.45 | 0.00 | 5.34 | 1.68 | 0.00 | 13.48 | 21.73 |
| | Maximum | 68.34 | 92.46 | 55.28 | 64.32 | 93.47 | 71.36 | 94.47 | 76.38 | 91.96 | 39.20 | 96.48 |
| | N | 5 | 5 | 5 | 5 | 5 | 5 | 5 | 5 | 5 | 5 | 200 |
| SCM | Minimum | 92.00 | 77.88 | 99.86 | 84.74 | 90.32 | 61.36 | 39.28 | 60.85 | 82.66 | 35.57 | 6.87 |
| | Average | 92.80 | 85.48 | 99.87 | 87.37 | 91.79 | 67.17 | 56.87 | 71.35 | 83.05 | 40.12 | 77.28 |
| | St. Dev. | 0.60 | 9.02 | 0.01 | 3.09 | 1.16 | 6.79 | 11.24 | 8.08 | 0.40 | 3.38 | 19.78 |
| | Maximum | 93.42 | 95.68 | 99.88 | 92.62 | 93.38 | 75.96 | 66.99 | 78.69 | 83.71 | 44.98 | 6.87 |
| | N | 5 | 5 | 5 | 5 | 5 | 5 | 5 | 5 | 5 | 5 | 200 |
| SPW | Minimum | 52.94 | 70.59 | 64.71 | 41.18 | 64.71 | 41.18 | 64.71 | 23.53 | 64.71 | 47.06 | 23.53 |
| | Average | 52.94 | 72.94 | 65.88 | 41.18 | 69.41 | 41.18 | 71.77 | 25.88 | 64.71 | 48.24 | 58.05 |
| | St. Dev. | 0.00 | 3.22 | 2.63 | 0.00 | 2.26 | 0.00 | 4.92 | 3.22 | 0.00 | 2.63 | 16.14 |
| | Maximum | 52.94 | 76.47 | 70.59 | 41.18 | 70.59 | 41.18 | 76.47 | 29.41 | 64.71 | 47.06 | 88.24 |
| | N | 5 | 5 | 5 | 5 | 5 | 5 | 5 | 5 | 5 | 5 | 200 |

(continued)

**Table 3** (continued)

| | Economy | Seychelles | Sierra Leone | South Africa | Sudan | Sao Tome | Tanzania | Togo | Uganda | Zambia | Zimbabwe | sample |
|---|---|---|---|---|---|---|---|---|---|---|---|---|
| STW | Minimum | 68.34 | 88.44 | 54.27 | 63.32 | 92.46 | 71.36 | 81.41 | 72.36 | 91.96 | 9.05 | 1.01 |
| | Average | 68.34 | 90.45 | 54.87 | 63.31 | 93.27 | 71.36 | 90.25 | 73.77 | 91.96 | 15.08 | 73.09 |
| | St. Dev. | 0.00 | 2.01 | 0.55 | 0.00 | 0.45 | 0.00 | 5.34 | 1.68 | 0.00 | 13.48 | 21.73 |
| | Maximum | 68.34 | 92.46 | 55.28 | 63.32 | 93.47 | 71.36 | 94.47 | 76.38 | 91.96 | 39.20 | 96.48 |
| | N | 5 | 5 | 5 | 5 | 5 | 5 | 5 | 5 | 5 | 5 | 200 |
| SCW | Minimum | 92.00 | 77.88 | 99.86 | 84.74 | 90.32 | 61.36 | 39.28 | 60.85 | 82.66 | 35.57 | 6.87 |
| | Average | 92.80 | 85.48 | 99.87 | 87.37 | 91.37 | 67.17 | 56.87 | 71.35 | 83.05 | 40.12 | 77.27 |
| | St. Dev. | 0.60 | 9.02 | 0.01 | 3.09 | 1.16 | 6.79 | 11.24 | 8.08 | 0.40 | 3.38 | 19.79 |
| | Maximum | 93.42 | 95.68 | 99.88 | 92.63 | 93.38 | 75.96 | 66.99 | 78.69 | 83.71 | 44.98 | 99.88 |
| | N | 5 | 5 | 5 | 5 | 5 | 5 | 5 | 5 | 5 | 5 | 200 |
| SPMC | Minimum | 100 | 97.73 | 100 | 100 | 29.71 | 100 | 2.87 | 100 | 100 | 100 | 2.87 |
| | Average | 100 | 99.16 | 100 | 100 | 40.38 | 100 | 73.69 | 100 | 100 | 100 | 89.50 |
| | St. Dev. | 0.00 | 1.16 | 0.00 | 0.00 | 8.41 | 0.00 | 39.59 | 0.00 | 0.00 | 0.00 | 21.64 |
| | Maximum | 100 | 100 | 100 | 100 | 51.90 | 100 | 92.11 | 100 | 100 | 100 | 100 |

We define a DMU as an economy considered in a given year. This resulted in 200 DMUs. Note that the DEA models that we will consider in this study have 1 input and 10 outputs. With 200 DMUs, all our DEA models fulfil the rule of thumb (triple-sum rule: $200 \geq 33 = (3 \div (1 + 10))$) for qualitatively good models (Sarkis, 2007).

To validate our proposed seven composite measures, we compare them with each other and with the SSB measures using correlation analysis. Our results are presented in the next section.

# 4 Results and Analysis

Using the panel data of 200 observations collected previously, we first computed the seven composite measures (ICE1, ICE2, ICE3, ICE4, ICE5, ICE5, ICE6, ICE7) described in the previous section for each of the 200 DMUs. Then, we performed correlation analysis among them and with comparison with SSB to validate them. Next, we obtained their descriptive statistics per year and per economies that we have compared to the ones from SSB. We also performed ANOVA analyses per year and per economy. The results are presented and analyzed below.

First, we present in Table 4 the eigenvalues and the percentage of variance explained by the principal components when we apply PCA to our data.

We can observe that the first principal component explained 51.66% of the variance. We had decided for this analysis to explore composite indexes based on the two extreme possibilities: only the first principal component considered, and all the principal components considered. However, in a more practical way, the first five components would have sufficed to incorporate 100% of the variations. A common rule of thumb to determine the number of principal components (PC) is to consider only those associated with eigenvalues that are greater than or equal to one (1) (Nicoletti et al., 2000). In our case here, we would have then selected only the first two principal components who correspond to about 78% of the variance explained.

Table 5 (respectively Table 6) presents the Pearson (respectively Spearman) correlation matrices of the seven proposed composite measures and SSB, with the

**Table 4** Total Variance Explained—PCA

| Component | Eigenvalues | % of Variance explained | Cumulative % of variance explained |
|---|---|---|---|
| 1 | 3.6160 | 51.66 | 51.66 |
| 2 | 1.8760 | 26.80 | 78.46 |
| 3 | 0.7690 | 10.99 | 89.45 |
| 4 | 0.7320 | 10.45 | 99.90 |
| 5 | 0.0070 | 0.10 | 100.0 |
| 6 | 0.0000 | 0.00 | 100.00 |
| 7 | 0.0000 | 0.00 | 100.00 |

**Table 5** Pearson Correlations between measures

|        |         | SSB | S1ICE1 | S1ICE2 | S1ICE3 | S1ICE4 | S1ICE5 | S1ICE6 |
|--------|---------|-----|--------|--------|--------|--------|--------|--------|
| S1ICE1 | Pearson | 0.635*** |        |        |        |        |        |        |
|        | Sig.    | 0.000 |        |        |        |        |        |        |
| S1ICE2 | Pearson | **0.934***** | 0.651*** |        |        |        |        |        |
|        | Sig.    | 0.000 | 0.000 |        |        |        |        |        |
| S1ICE3 | Pearson | 0.667*** | 0.338*** | 0.524*** |        |        |        |        |
|        | Sig.    | 0.000 | 0.000 | 0.000 |        |        |        |        |
| S1ICE4 | Pearson | **0.911***** | 0.565*** | **0.818***** | **0.906***** |        |        |        |
|        | Sig.    | 0.000 | 0.000 | 0.000 | 0.000 |        |        |        |
| S1ICE5 | Pearson | 0.667*** | 0.339*** | 0.525*** | **1.000***** | **0.907***** |        |        |
|        | Sig.    | 0.000 | 0.000 | 0.000 | 0.000 | 0.000 |        |        |
| S1ICE6 | Pearson | 0.657*** | 0.324*** | 0.514*** | **1.000***** | **0.900***** | **1.000***** |        |
|        | Sig.    | 0.000 | 0.000 | 0.000 | 0.000 | 0.000 | 0.000 |        |
| S1ICE7 | Pearson | **0.717***** | 0.166* | 0.477*** | 0.678*** | **0.701**** | 0.678*** | 0.679*** |
|        | Sig.    | 0.000 | 0.019 | 0.000 | 0.000 | 0.000 | 0.000 | 0.000 |

*, **, ***: Significant at 10%, 5%, and 1% respectively

corresponding significance. Note that in both tables, strong correlation coefficients (greater than 0.70) are in bold.

In Table 5, one can observe that the Pearson correlation coefficient of each of our proposed seven new measures with SSB is strong, or almost strong (greater than 0.63), and they are all significant. In fact, all the Pearson correlation coefficients between the eight measures are significant. The strongest Pearson correlations of SSB are with ICE2, ICE4, and ICE7, respectively, in decreasing order. As once could have expected, all our proposed DEA-based measures are strongly, or almost strongly correlated (greater than 0.67) to each other, with a couple of perfect correlations. ICE1 and ICE2, both based on PCA, are almost strongly correlated. Finally, when we compared PCA-based measures with DEA-based ones, the only strong correlation is between ICE2 and ICE4.

With the Spearman's correlation coefficients, the observations are slightly different. Here we have some correlation coefficients that are not significant, and there is no perfect correlation. When comparing SSB with the seven proposed measures, all the Spearman coefficients are significant, as were the Pearson ones. However, only three are strong and one is almost strong. Interestingly, as with the Pearson ones, the strongest correlation is with ICE2, followed in that order by ICE4, and ICE7. The correlation between the two PCA-based measure is strong and significant. Between DEA-based measures, all correlations are significant, and they are either strong (ICE3 and ICE5, ICE3 and ICE6, ICE5 and ICE6), average, or low (ICE4 and 6). When one compares PCA-based and DEA-based measures here, we again see a difference with the Pearson coefficients values. Correlations are average, weak, and even negative in most cases. In general, correlations with ICE1 are weaker than correlations with ICE2. And there are only two strong correlations (ICE2 and ICE4, and ICE2 and ICE7).

**Table 6** Spearman Correlations between measures

| | | SSB | S1ICE1 | S1ICE2 | S1ICE3 | S1ICE4 | S1ICE5 | S1ICE6 |
|---|---|---|---|---|---|---|---|---|
| S1ICE1 | Spearman | 0.690*** | | | | | | |
| | Sig. | 0.000 | | | | | | |
| S1ICE2 | Spearman | 0.927*** | 0.759*** | | | | | |
| | Sig. | 0.000 | 0.000 | | | | | |
| S1ICE3 | Spearman | 0.397*** | 0.069 | 0.241 | | | | |
| | Sig. | 0.000 | 0.332 | 0.001 | | | | |
| S1ICE4 | Spearman | 0.967*** | 0.670*** | 0.891*** | 0.518*** | | | |
| | Sig. | 0.000 | 0.000 | 0.000 | 0.000 | | | |
| S1ICE5 | Spearman | 0.386*** | 0.091*** | 0.234*** | 0.963*** | 0.503*** | | |
| | Sig. | 0.000 | 0.201 | 0.001 | 0.000 | 0.000 | | |
| S1ICE6 | Spearman | 0.056 | −0.257*** | −0.070 | 0.864*** | 0.161** | 0.822*** | |
| | Sig. | 0.434 | 0.000 | 0.324 | 0.000 | 0.023 | 0.000 | |
| S1ICE7 | Spearman | 0.900*** | 0.469*** | 0.781*** | 0.554*** | 0.887*** | 0.537*** | 0.302*** |
| | Sig. | 0.000 | 0.000 | 0.000 | 0.000 | 0.000 | 0.000 | 0.000 |

*, **, ***: Significant at 10%, 5%, and 1% respectively

**Table 7** Averages—ANOVA by year

| Year | SSB | ICE1 | ICE2 | IEC3 | IEC4 | IEC5 | IEC6 | IEC7 |
|------|-----|------|------|------|------|------|------|------|
| 2014 | 64.34 | 56.11 | 69.40 | 94.42 | 87.12 | 94.43 | 94.26 | 76.92 |
| 2015 | 67.52 | 56.99 | 72.19 | 95.51 | 88.91 | 95.51 | 95.34 | 82.39 |
| 2016 | 72.14 | 56.99 | 73.98 | 96.08 | 89.99 | 96.09 | 95.92 | 87.04 |
| 2017 | 74.52 | 59.20 | 75.52 | 97.23 | 91.47 | 97.23 | 97.04 | 89.80 |
| 2018 | 76.67 | 60.96 | 77.08 | 98.96 | 93.19 | 98.98 | 98.77 | 92.20 |
| Sample | 71.04 | 58.05 | 73.63 | 96.44 | 90.14 | 96.45 | 96.27 | 85.67 |
| ANOVA | | | | | | | | |
| $F$-value | **4.775** | 0.602 | 1.623 | 1.884 | **2.751** | 1.894 | 1.855 | **5.658** |
| $P$-value | **0.001**[***] | 0.662 | 0.170 | 0.115 | **0.029**[**] | 0.113 | 0.120 | **0.000**[***] |

*, **, ***: Significant at 10%, 5%, and 1% respectively

In summary, all our proposed seven measures are valid, when compared to SSB. We obtain measures that are mostly significant and strongly correlated with the SSB measure of Doing Business. Between the two PCA-based measures, the ICE2 appears the strongest for both performance assessment and ranking purpose. Among the DEA-based measures, ICE4 and ICE7 appear the strongest for the same purpose.

In our next analysis, we computed annual averages of the seven proposed measures and contrasted in with the one from SSB. We have also performed ANOVA analyses. Table 7 presents the annual averages, then the $F$-value, and $P$-vales of the ANOVA analysis.

A graphical visualization of the annual averages is provided in Fig. 1.

From both Table 7 and Fig. 1, all the seven new composite measures confirm that there has been some steady improvement in the performances of the sample regarding how the regulatory environment impacts the creation of businesses. Where the measures disagree is on the magnitude of the year-to-year increases. The ANOVA analysis confirms this observation and shows that it is significant for three measures: SSB, and the two DEA-based measures it had strong correlations with: ICE4 and ICE7.

In our subsequent analysis, we computed averages by economies of the seven proposed measures and contrasted in with the ones from SSB. We also performed ANOVA analyses. Table 8 presents the averages, then the $F$-value, and $P$-values of the ANOVA analysis. Table 9 reports the respective ranks of each country, according to the average measures. Finally, Fig. 2 presents a visualization of the eight composite measures when the countries are ranked according to SSB.

From Table 8, we can observe the drawback of the classic CCR models that we anticipated due to the fact that several countries, in average, had the maximum possible value for one sub-indicator, SPMC. Indeed, we have countries with an average composite IEC3 score of 100. The three post hoc DEA models that we used to correct this drawback, DEA with weights restrictions (IEC4), Bootstrap DEA (IEC6) and Benevolent Cross-Efficiency (IEC7) did rectify the issue and were able to increase discrimination among the countries. Interestingly, the super-efficiency

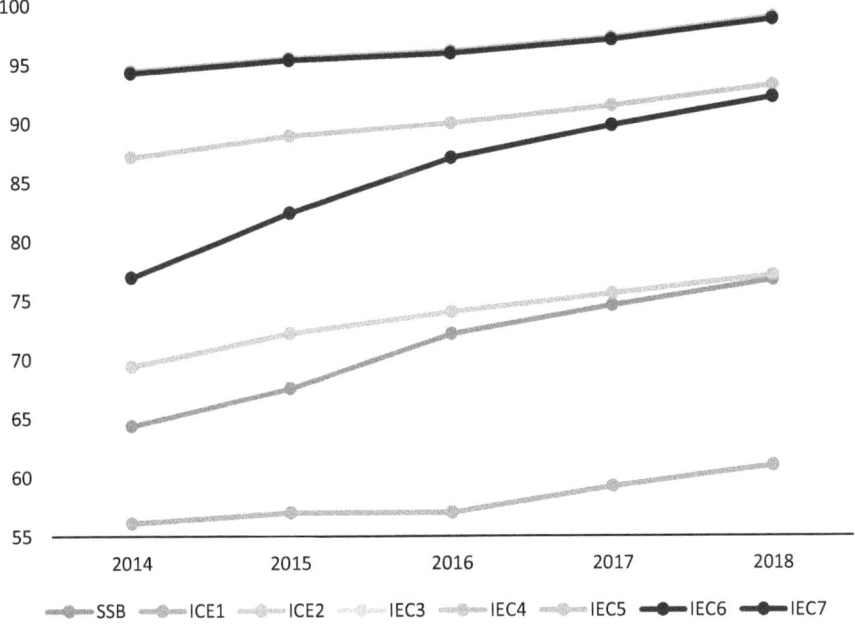

**Fig. 1** Annual averages for all measures

DEA (IEC5) did not really increase discrimination between countries. There is hence a necessity to perform a further analysis at the DMU level to further understand the dynamic at play.

The ANOVA analyses by country, which results are presented in Table 8, show that, in average, there are significant differences between the performances of the 40 countries of the sample according to all the eights measures.

To conclude our analyses, given that all the observations above provide a validation of the proposed seven (7) measures, we have performed a ranking of the 40 countries, given that it was one of the reasons we needed performance measures and presents those ranks in Table 9. In the table, the top 10 ranked countries are highlighted.

The number one ranked country, according to SSB, Burundi, holds the same ranks for four of the seven proposed measures (ICE1, IEC2, IEC3, IEC7), is second ranked for IEC4, and third for IEC6. There is hence a clear and strong majority of the measures to identify Burundi as one of the best performers of the sample. In addition, the top three countries, according to SSB, Burundi, Mauritius, and Liberia, are in the top three for five of the seven proposed measures (IEC2, IEC3, IEC4, IEC5, IEC7). Note that in all the above cases, measures using the two data-driven approaches (PCA and DEA) agree with each other. The overall agreement between the measures when it comes to ranking the 40 countries according to their average performance is further illustrated in Fig. 2.

**Table 8** Averages—ANOVA by countries

| Country | SSB | ICE1 | ICE2 | IEC3 | IEC4 | IEC5 | IEC6 | IEC7 |
|---|---|---|---|---|---|---|---|---|
| Angola | 69.71 | 60.00 | 62.32 | 96.97 | 88.07 | 96.97 | 96.92 | 87.54 |
| Benin | 79.36 | 63.53 | 81.79 | 96.08 | 92.26 | 96.07 | 95.81 | 85.94 |
| Botswana | 76.56 | 52.94 | 70.25 | **100.0** | **93.69** | **100.0** | **99.75** | **90.93** |
| Burkina Faso | 80.24 | 88.24 | 83.13 | **100.0** | **95.49** | **100.0** | **99.70** | **73.21** |
| **Burundi** | 93.78 | 88.24 | 93.88 | **100.0** | **99.70** | **100.0** | **99.18** | **99.66** |
| Cabo Verde | 82.02 | 52.94 | 82.47 | **100.0** | **94.89** | **100.0** | **99.93** | **95.36** |
| Cameroon | 75.78 | 64.71 | 79.75 | 89.34 | 86.97 | 89.34 | 88.96 | 72.83 |
| Chad | 44.04 | 52.94 | 38.59 | 73.59 | 63.28 | 73.59 | 73.50 | 58.95 |
| Comoros | 66.12 | 52.94 | 70.74 | 90.24 | 83.26 | 90.24 | 90.15 | 73.21 |
| Congo, Rep. | 58.38 | 36.47 | 52.32 | 83.41 | 75.05 | 83.41 | 83.38 | 74.35 |
| Côte d'Ivoire | 86.92 | 75.29 | 88.94 | 98.53 | 96.35 | 98.53 | 98.34 | 91.67 |
| Eritrea | 46.15 | 29.41 | 39.31 | 80.63 | 70.37 | 80.63 | 80.43 | 51.60 |
| Eswatini | 73.30 | 35.29 | 72.24 | 99.92 | 91.28 | 99.92 | 99.88 | 91.83 |
| Ethiopia | 54.77 | 25.88 | 57.74 | 72.88 | 67.22 | 72.88 | 72.69 | 66.56 |
| Gabon | 75.20 | 54.11 | 69.75 | 96.82 | 90.66 | 96.82 | 96.71 | 88.77 |
| Gambia, The | 67.07 | 63.53 | 63.59 | **100.0** | **88.78** | **100.0** | **99.96** | **91.97** |
| Ghana | 83.68 | 58.82 | 84.55 | 99.36 | 95.18 | 99.37 | 99.29 | 95.79 |
| Kenya | 73.52 | 41.18 | 72.61 | **100.0** | **91.42** | **100.0** | **99.97** | **92.48** |
| Lesotho | 82.87 | 64.71 | 79.52 | **100.0** | **95.54** | **100.0** | **99.90** | **94.37** |
| Liberia | 90.59 | 76.47 | 91.52 | **100.0** | **98.40** | **100.0** | **99.82** | **98.71** |
| Madagascar | 81.38 | 57.65 | 82.72 | **100.0** | **94.63** | **100.0** | **99.94** | **96.12** |
| Malawi | 70.33 | 58.82 | 64.85 | **100.0** | **90.11** | **100.0** | **99.97** | **90.97** |
| Mali | 72.07 | 76.47 | 78.40 | 95.57 | 90.04 | 95.57 | 95.38 | 64.06 |
| Mauritania | 77.93 | 65.88 | 83.93 | 96.05 | 92.19 | 96.05 | 95.80 | 73.20 |
| **Mauritius** | 91.66 | 70.59 | 92.94 | **100.0** | **99.77** | **100.18** | **99.36** | **98.90** |
| Mozambique | 74.03 | 47.06 | 74.16 | **100.0** | **91.62** | **100.0** | **99.97** | **93.59** |

| | | | | | | | | |
|---|---|---|---|---|---|---|---|---|
| Namibia | 69.37 | 47.06 | 59.47 | 100.0 | 89.75 | 100.0 | 99.95 | 87.51 |
| Nigeria | 77.79 | 56.12 | 75.59 | 100.0 | 93.16 | 100.0 | 99.97 | 93.58 |
| Rwanda | 84.81 | 68.24 | 86.42 | 100.0 | 96.03 | 100.0 | 99.76 | 97.60 |
| São Tomé and Príncipe | 73.71 | 69.41 | 84.26 | 97.07 | 92.66 | 97.07 | 96.63 | 53.69 |
| Senegal | 84.19 | 82.35 | 85.92 | 98.47 | 95.30 | 98.47 | 98.13 | 89.79 |
| Seychelles | 78.52 | 52.94 | 75.61 | 100.0 | 93.46 | 100.0 | 99.95 | 93.05 |
| Sierra Leone | 87.01 | 72.94 | 87.36 | 99.16 | 96.54 | 99.16 | 98.99 | 96.94 |
| South Africa | 80.16 | 65.88 | 72.83 | 100.0 | 95.35 | 100.07 | 99.48 | 91.92 |
| Sudan | 73.83 | 41.18 | 70.64 | 100.0 | 91.54 | 100.0 | 99.97 | 91.61 |
| Tanzania | 69.93 | 41.18 | 68.51 | 100.0 | 89.95 | 100.0 | 99.98 | 91.70 |
| Togo | 73.14 | 71.76 | 77.02 | 93.54 | 88.19 | 93.54 | 93.29 | 76.61 |
| Uganda | 67.75 | 25.88 | 68.54 | 100.0 | 89.06 | 100.0 | 99.97 | 91.36 |
| Zambia | 84.93 | 64.71 | 86.31 | 100.0 | 96.08 | 100.0 | 99.93 | 97.22 |
| Zimbabwe | 50.86 | 48.24 | 34.82 | 100.0 | 82.15 | 100.0 | 99.98 | 81.66 |
| ANOVA | | | | | | | | |
| F-value | 14.066 | 62.391 | 39.898 | 11.707 | 17.609 | 11.715 | 11.613 | 5.356 |
| P-value | 0.000*** | 0.000*** | 0.000*** | 0.000*** | 0.000*** | 0.000*** | 0.000*** | 0.000*** |

**Table 9** Country ranks according to average measures

| Country | SSB | ICE1 | ICE2 | IEC3 | IEC4 | IEC5 | IEC6 | IEC7 |
|---|---|---|---|---|---|---|---|---|
| Angola | 31 | 19 | 34 | 29 | 33 | 29 | 28 | 26 |
| Benin | 15 | 17 | 15 | 31 | 19 | 31 | 31 | 28 |
| Botswana | 19 | 25 | 28 | 1 | 15 | 3 | 19 | 23 |
| Burkina Faso | 13 | 1 | 12 | 1 | 9 | 3 | 20 | 32 |
| **Burundi** | **1** | **1** | **1** | **1** | **2** | **3** | 24 | **1** |
| Cabo Verde | 11 | 25 | 14 | 1 | 13 | 3 | 13 | 9 |
| Cameroon | 20 | 14 | 16 | 36 | 34 | 36 | 36 | 35 |
| Chad | 40 | 25 | 39 | 39 | 40 | 39 | 39 | 38 |
| Comoros | 35 | 25 | 26 | 35 | 35 | 35 | 35 | 32 |
| Congo, Rep. | 36 | 36 | 37 | 37 | 37 | 37 | 37 | 31 |
| Côte d'Ivoire | **5** | 6 | **4** | 26 | **5** | 26 | 26 | 19 |
| Eritrea | 39 | 38 | 38 | 38 | 38 | 38 | 38 | 40 |
| Eswatini | 26 | 37 | 25 | 23 | 24 | 23 | 16 | 17 |
| Ethiopia | 37 | 39 | 36 | 40 | 39 | 40 | 40 | 36 |
| Gabon | 21 | 24 | 29 | 30 | 25 | 30 | 29 | 25 |
| Gambia, The | 34 | 17 | 33 | 1 | 31 | 3 | 9 | 15 |
| Ghana | 9 | 20 | 9 | 24 | 12 | 24 | 23 | 8 |
| Kenya | 25 | 33 | 24 | 1 | 23 | 3 | 3 | 14 |
| Lesotho | 10 | 14 | 17 | 1 | 8 | 3 | 15 | 10 |
| Liberia | **3** | **4** | **3** | 1 | **3** | 3 | 17 | **3** |
| Madagascar | 12 | 22 | 13 | 1 | 14 | 3 | 12 | 7 |
| Malawi | 29 | 20 | 32 | 1 | 26 | 3 | 3 | 22 |
| Mali | 28 | **4** | 18 | 33 | 27 | 33 | 33 | 37 |
| Mauritania | 17 | 12 | 11 | 32 | 20 | 32 | 32 | 34 |
| **Mauritius** | **2** | 9 | **2** | **1** | **1** | **1** | 22 | **2** |
| Mozambique | 22 | 31 | 22 | 1 | 21 | 3 | 3 | 11 |
| Namibia | 32 | 31 | 35 | 1 | 29 | 3 | 10 | 27 |
| Nigeria | 18 | 23 | 21 | 1 | 17 | 3 | 3 | 12 |
| Rwanda | 7 | 11 | 6 | 1 | 7 | 3 | 18 | **4** |
| São Tomé and Príncipe | 24 | 10 | 10 | 28 | 18 | 28 | 30 | 39 |
| Senegal | 8 | **3** | 8 | 27 | 11 | 27 | 27 | 24 |
| Seychelles | 16 | 25 | 20 | 1 | 16 | 3 | 10 | 13 |
| Sierra Leone | **4** | 7 | **5** | 25 | **4** | 25 | 25 | 6 |
| South Africa | 14 | 12 | 23 | 1 | 10 | **2** | 21 | 16 |
| Sudan | 23 | 33 | 27 | 1 | 22 | 3 | 3 | 20 |
| Tanzania | 30 | 33 | 31 | 1 | 28 | 3 | 1 | 18 |
| Togo | 27 | 8 | 19 | 34 | 32 | 34 | 34 | 30 |
| Uganda | 33 | 39 | 30 | 1 | 30 | 3 | 3 | 21 |
| Zambia | 6 | 14 | 7 | 1 | 6 | 3 | 13 | 5 |
| Zimbabwe | 38 | 30 | 40 | 1 | 36 | 3 | 1 | 29 |

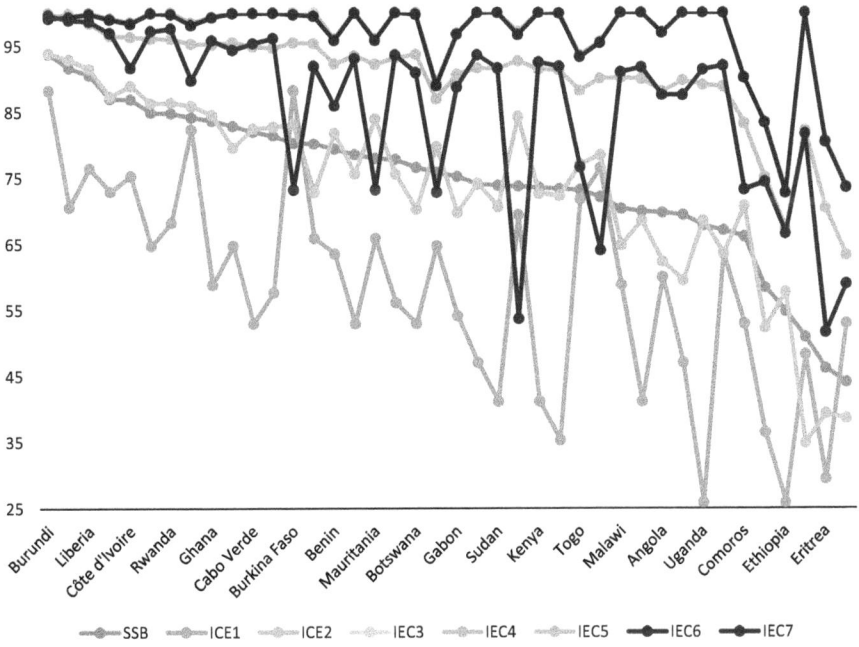

**Fig. 2** Country averages for all measures—Country ranked according to SSB

# 5 Limitations of the Case Study

The objective of this case study was to validate data-driven tools from operations research, Principal Component Analysis (PCA) and Data Envelopment Analysis (DEA), to construct composite indexes to assess how the regulatory ecosystem promotes entrepreneurial activities. In the previous sections, we have achieved in part that goal by proposing and validating seven (7) alternative measures of the impact of regulatory activities on one fundamental dimension of entrepreneurial activities. Hence, our work comes with limitations that we need to acknowledge, mainly on scope, methodologies and their added value, and validity of the data that we have used.

First, we only analyzed one dimension of the entrepreneurial activities, and not the entire ecosystem. Our conclusions are therefore limited in their scope. In addition, only a 5 year period is considered. To strengthen our observations, we will need to improve this work in two directions. One would be to extend the PCA and DEA approaches to the nine other dimensions of the ecosystem. The other would be, in the second phase, to construct an overall composite measure using again either PCA or DEA, or a combination of the two.

Regarding the methodologies, when using PCA, we decided a priori the number of principal components to consider in the composite index, leading to either using the first one, hence only considering about half of the variance explained; or using all of them, only to realize that we considered way more PCs than we needed. The methodology has to be improved by letting the data guide the selection on the number of principal components. In addition, we have only considered average performance at the country level over the study period. Doing so has prevented us from performing a more detailed analysis at the micro-level, the DMU level (a country considered a given year). As a result, we have also not leveraged the added value of using DEA models. Indeed, in addition to rankings, using DEA allows us to perform a benchmarking at the DMU level: identify which DMU (economy a given year) exhibits the best practices (efficient ones), and for those who do not have best practices, which efficient DMUs they should emulate to improve their performance and become efficient. Consequently, we could identify policy insights that could guide adjustments in regulations to improve countries standings. Further, we noticed that the Super-Efficiency DEA models did not significantly enhance the discrimination between the efficiency units. There is a need for more analyses to understand the dynamic at play.

Finally, there was an unforeseen issue with the validity of the data that were used in this study. In September 2021, months after this study has started, the World Bank announced that it was discontinuing its Doing Business Project after data irregularities in its latest reports were reported (World Bank, 2020, 2021, 2022). Fortunately, according to the audit report (World Bank, 2020), the countries involved in our case studies were not concerned with the irregularities. Nevertheless, our intent in the future would be to update our analysis using the curated data that the World Bank is currently working on, use existing alternate indicators if they exist, or further validate our composite measures through regression analysis using indicators or variables known to impact entrepreneurial activities.

## 6   Conclusion and Future Projects

In summary, we presented a case study on applying data-driven approaches to construct composite scores aimed at evaluating how the regulatory ecosystem promotes entrepreneurial activities, more specifically, the start a business dimension. We proposed seven (7) new composite measures. The first two are based on Principal Component Analysis and use a pre-determined number of principal components (the first and all). The remaining five (5) are based on Data Envelopment Analysis (DEA) models: the classical CCR, the CCR with weights restrictions, the super-efficiency CCR, the Bootstrap CCR, and the benevolent cross-efficiency CCR. The proposed composite measures are compared to the Start a Business score published by the Doing Business Project of the World Bank and validated using correlation and consistency analysis. We used a sample of 40 sub-Saharan economies during the 5 year period (2014–2018).

From our analysis, the proposed measures are consistent with the SSB scores. We used them to determine the annual average scores of the sample, and the average score of each country in our sample and compared our results with the one published by Doing Business. We observed a consensus on all measures on the fact that overall, the performance of the regulatory environment to promote the start of business have improved steadily every year. Our ANOVA analyses confirmed it, significantly for two measures, SSB and the benevolent cross-efficiency DEA model. Regarding the average score by country, there was consensus from all measures that in average, there were differences between the 40 countries. The ANOVA analyses were significant for all the measures considered. Finally, we ranked the countries according to their average scores. A majority of the measures, always including measures based on both the PCA and the DEA approaches, agrees on the country with the best performance and the ones in the top three performances in our sample.

Nevertheless, there were some limitations to our work, as described in the previous section. The limitations were related to the scope of the case study, the methodologies and approaches used, and the validity of the data used. We have identified as well, how one can remedy to these limitations.

# Bibliography

Adusei, M. (2016). Does entrepreneurship promote economic growth in Africa? *African Development Review, 28*(2), 201–214.

Ahamed, M. M., & Mallick, S. K. (2019). Is financial inclusion good for bank stability? International evidence. *Journal of Economic Behaviour & Organisation, 157*(1), 403–427.

Ajide, F. M. (2020). Infrastructure and Entrepreneurship: Evidence from Africa. *Journal of Developmental Entrepreneurship, 25*(3), 2050015. https://doi.org/10.1142/S1084946720500156

Álvarez, I. C., Barbero, J., & Zofío, J. L. (2020). A data envelopment analysis toolbox for MATLAB. *Journal of Statistical Software, 95*, 1–49.

Anarfo, E. B., Abor, J. Y., & Osei, K. A. (2020). Financial regulation and financial inclusion in Sub-Saharan Africa: Does financial stability play a moderating role? *Research in International Business and Finance, 51*(1), 101–070.

Angulo-Meza, L., & Lins, M. P. E. (2002). Review of methods for increasing discrimination in data envelopment analysis. *Annals of Operations Research, 116*(1), 225–242.

Assaf, A. G., Barros, C., & Sellers-Rubio, R. (2011). Efficiency determinants in retail stores: a Bayesian framework. *Omega, 39*(1), 283–292.

Ayyagari, M., Demirguc-Kunt, A., and Maksimovic, V. (2011). *Small vs. young firms across the world: Contribution to employment, job creation, and growth.* Policy Research Working Paper WPS 5631; p. 59. World Bank. http://hdl.handle.net/10986/3397

Banker, R. D., Charnes, A., & Cooper, W. W. (1984). Some models for estimating technical and scale efficiencies in data envelopment analysis. *Management Science, 30*(9), 1078–1092.

Banker, R. D., Charnes, A., Cooper, W. W., & Maindiratta, A. (1988). A comparison of DEA and translog estimates of production frontiers using simulated observations from a known technology. In *Applications of modern production theory: Efficiency and productivity* (pp. 33–55). Springer .

Banker, R. D., Conrad, R. F., & Strauss, R. P. (1986). A comparative application of data envelopment analysis and translog methods: An illustrative study of hospital production. *Management Science, 32*(1), 30–44.

Becker, W., Saisana, M., Paruolo, P., & Vandecasteele, I. (2017). Weights and importance in composite indicators: Closing the gap. *Ecological Indicators, 80,* 12–22. https://doi.org/10.1016/j.ecolind.2017.03.056

Ben Youssef, A., Boubaker, S., Dedaj, B., & Carabregu-Vokshi, M. (2021). Digitalization of the economy and entrepreneurship intention. *Technological Forecasting and Social Change, 164,* 120043. https://doi.org/10.1016/j.techfore.2020.120043

Cámara, N., and Tuesta, D. (2014). Measuring financial inclusion: A multidimensional index. *BBVA Research Paper* (14/26).

Charnes, A., Cooper, W. W., & Rhodes, E. (1978). Measuring the efficiency of decision-making units. *European Journal of Operational Research, 2*(6), 429–444.

Chen, Y., Cook, W. D., & Lim, S. (2019). Preface: DEA and its applications in operations and data analytics. *Annals of Operations Research, 278*(1–2), 1–4.

Cherchye, L. (2001). Using data envelopment analysis to assess macroeconomic policy performance. *Applied Economics, 33*(3), 407–416.

Cherchye, L., & Kuosmanen, T. (2004). *Benchmarking sustainable development: A synthetic meta-index approach (No. 2004/28).* UNU-WIDER research paper, United Nations University (UNU).

Cherchye, L., Moesen, W., & Van Puyenbroeck, T. (2004). Legitimately diverse, yet comparable: On synthesizing social inclusion performance in the EU. *JCMS Journal of Common Market Studies, 42*(5), 919–955.

Daraio, C., & Simar, L. (2007). The measurement of efficiency. In C. Daraio & L. Simar (Eds.), *Advanced robust and nonparametric methods in efficiency analysis.* Springer.

Djankov, S. (2016). The doing business project: How it started: Correspondence. *Journal of Economic Perspectives, 30*(1), 247–248.

Doyle, J., & Green, R. (1994). Efficiency and cross-efficiency in DEA: Derivations, meanings and uses. *Journal of the Operational Research Society, 45*(5), 567–578.

Dutta, N., & Sobel, R. (2016). Does corruption ever help entrepreneurship? *Small Business Economics, 47*(1), 179–199. https://doi.org/10.1007/s11187-016-9728-7

Emrouznejad, A., & Yang, G. L. (2018). A survey and analysis of the first 40 years of scholarly literature in DEA: 1978–2016. *Socio-Economic Planning Sciences, 61*(1), 4–8.

Fossen, F. M., & Sorgner, A. (2019). Digitalization of work and entry into entrepreneurship. *Journal of Business Research, 125,* 548–563. https://doi.org/10.1016/j.jbusres.2019.09.019

Foster, J. E., McGillivray, M., & Seth, S. (2013). Composite indices: Rank robustness, statistical association, and redundancy. *Econometric Reviews, 32*(1), 35–56. https://doi.org/10.1080/07474938.2012.690647

Gedeon, S. (2010). What is entrepreneurship? *Entrepreneurial Practice Review, 1*(3), 21.

Greco, S., Ishizaka, A., Tasiou, M., & Torrisi, G. (2019). On the methodological framework of composite indices: A review of the issues of weighting, aggregation, and robustness. *Social Indicators Research, 141*(1), 61–94.

Greyling, T., & Tregenna, F. (2017). Construction and analysis of a composite quality of life index for a region of South Africa. *Social Indicators Research, 131*(3), 887–930.

IBM. (2016). *IBM SPSS Statistics for windows, version 24.0.* IBM Corp.

International Labour Office. (2015). *Small and medium-sized enterprises and decent and productive employment creation: Fourth item on the agenda.* International Labour Office.

Lee, C.-C., Wang, C.-W., & Ho, S.-J. (2020). Country governance, corruption, and the likelihood of firms' innovation. *Economic Modelling, 92,* 326–338. https://doi.org/10.1016/j.econmod.2020.01.013

Lovell, C. K., Pastor, J. T., & Turner, J. A. (1995). Measuring macroeconomic performance in the OECD: A comparison of European and non-European countries. *European Journal of Operational Research, 87*(3), 507–518.

Mahlberg, B., & Obersteiner, M. (2001). *Remeasuring the HDI by data envelopment analysis. SSRN.* https://papers.ssrn.com/sol3/papers.cfm?abstract_id=1999372

Melyn, W., & Moesen, W. (1991). Towards a synthetic indicator of macroeconomic performance: Unequal weighting when limited information is available. *Public economics research papers*, 1–24.

Méndez-Picazo, M.-T., Galindo-Martín, M.-Á., & Ribeiro-Soriano, D. (2012). Governance, entrepreneurship and economic growth. *Entrepreneurship & Regional Development, 24*(9–10), 865–877. https://doi.org/10.1080/08985626.2012.742323

Meyer, N., & de Jongh, J. (2018). The importance of entrepreneurship as a contributing factor to economic growth and development: The case of selected European countries. *Journal of Economics and Behavioral Studies, 10*(4 (J)), 287–299.

Nardo, M., Saisana, M., Saltelli, A., & Tarantola, S. (2005). Tools for composite indicators building. *European Commission, Ispra, 15*(1), 19–20.

Nicoletti, G., Scarpetta, S., and Boylaud, O. (2000). *Summary indicators of product and market regulation with an extension to employment protection legislation.* Accessed March 19, 2022 www.oecd.org/eco/eco

Omri, A. (2020). Formal versus informal entrepreneurship in emerging economies: The roles of governance and the financial sector. *Journal of Business Research, 108*, 277–290. https://doi.org/10.1016/j.jbusres.2019.11.027

Ouattara, A., Takouda, P. M., and Dia, M. (2021)*UN Indice Agrégé D'inclusion Financière Base Sur L'analyse PAR Enveloppement De Donnees (DEA) En Contexte UEMOA. CESAG Working Papers—Spécial JRI, No 001/2021*, p. 88–95. https://www.cesag.sn/images/CESAG_WORKING_PAPERS-A_Ouattara_et_al.pdf

Permanyer, I. (2011). Assessing the robustness of composite indices rankings. *Review of Income and Wealth, 57*(2), 306–326. https://doi.org/10.1111/j.1475-4991.2011.00442.x

R Core Team. (2017). *R: A language and environment for statistical computing.* R Foundation for Statistical Computing. https://www.R-project.org/

Robbins, S. P., Coulter, M., and DeCenzo, D. A. (2020). *Fundamentals of management: Management myths debunked!* (Eleventh edition, global edition). Pearson.

Robson, P. J., Haugh, H. M., & Obeng, B. A. (2009). Entrepreneurship and innovation in Ghana: Enterprising Africa. *Small Business Economics, 32*(3), 331–350.

Rogge, N., & Kolyaseva, A. (2022). Measuring and comparing World Bank region's'ease of doing business' opportunity sets. *Journal of Productivity Analysis*, 1–25.

Sarkis, J. (2007). Preparing your data for DEA. In *Modeling data irregularities and structural complexities in data envelopment analysis* (pp. 305–320). Springer, .

Scheel, H. (2000). EMS: Efficiency measurement system user's manual. http://www.holger-scheel.de/ems/ems.pdf

Sergi, B. S., Popkova, E. G., Bogoviz, A. V., & Ragulina, J. V. (2019). Entrepreneurship and economic growth: The experience of developed and developing countries. In *Entrepreneurship and Development in the 21st Century*. Emerald.

Simar, L., & Wilson, P. W. (1998). Sensitivity analysis of efficiency scores: How to bootstrap in nonparametric frontier models. *Management Science, 44*(1), 49–61.

Simar, L., & Wilson, P. W. (2000). Statistical inference in nonparametric frontier models: The state of the art. *Journal of Productivity Analysis, 13*(1), 49–78.

Storrie, D., and Bjurek, H. (2000). *Benchmarking European labour market performance with efficiency frontier techniques (No. FS I 00–211)*. WZB Discussion Paper.

Takouda, P. M., Dia, M., and Ouattara, A. (2020, November). Levels of financial inclusion in the WAEMU countries: A case study using DEA. In *2020 International Conference on Decision Aid Sciences and Application (DASA)* (pp. 1274–1278). IEEE.

Toma, P., Miglietta, P. P., Zurlini, G., Valente, D., & Petrosillo, I. (2017). A non-parametric bootstrap-data envelopment analysis approach for environmental policy planning and management of agricultural efficiency in EU countries. *Ecological Indicators, 83*(1), 132–143.

Wamba, S. F., Gunasekaran, A., Dubey, R., & Ngai, E. W. (2018). Big data analytics in operations and supply chain management. *Annals of Operations Research, 270*(1–2), 1–4.

Wang, X., & Wang, Y.-M. (2021). Interval enhanced Russell measure with undesirable outputs based on data envelopment analysis: An efficiency measurement of industry in China. *Journal of Intelligent & Fuzzy Systems, 40*(1), 103–115. https://doi.org/10.3233/JIFS-182943

World Bank. (2011). *Doing business dans les Etats membres de l'OHADA 2012*. World Bank Group.

World Bank. (2019). *Ease of doing business score and ease of doing business ranking*. World Bank Group.

World Bank. (2020). *Management review of data irregularities in the doing business reports from 2016 to 2020: Verification report (English)*. World Bank Group.

World Bank. (2022). *Doing business*. World Bank. Accessed March 19, 2022, from https://archive.doingbusiness.org/en/doingbusiness

World Bank Group. (2021). *World Bank group to discontinue doing business report*. World Bank. Accessed March 19, 2022, from https://www.worldbank.org/en/news/statement/2021/09/16/world-bank-group-to-discontinue-doing-business-report

# Exact Resolution for Healthcare Facility Location Problem: A Case Study of the Specific Pharmacy of Tunisia

Fatma Ben Amor, Manel Kammoun, and Taicir Loukil

**Abstract** Healthcare facilities are important to all communities, for that their location need a strategic planning to select its best locations. In this context, we hope to focus on the problem of the location of the specific pharmacy of Tunisia. As its name indicates, this pharmacy is responsible for distributing the specific treatment to all the patients in all the Tunisian cities. We can imagine the congestion because of the assembly of large number of patients, who come to this pharmacy to take their treatments. In this paper, we are interested to solve this real case by involving a set of pharmacies to raise the total accessibility for the entire population. By applying this proposition, the demand of the patients of each city could be satisfied directly by visiting the city or indirectly by covering it. Experiments were made using real-world instances with multiple scenarios proposed.

## 1 Introduction

Given the growth of the population in Tunisia and therefore the number of patients is expanding, so new healthcare facilities are needed through the time. Since they are essential, so their location is an important issue in many cities planning. Many studies have long been interested to the problem of where should be located these facilities. This decision needs a strategic planning to select the best location. If the location is not well chosen, it has a lot of consequences such decreasing the accessibility for the entire population or reducing the coverage range.

In the context of health care services, we have the medicine markets, which are named pharmacies. The pharmacies are important in providing medicines and advices to the patients in case of any illness. When talking about diseases in Tunisia, some of them needs specific treatments such as Chemotherapy, Osteoporosis, and Parkinson's disease. The specific treatments are too expensive and indicated for the

F. Ben Amor (✉) · M. Kammoun · T. Loukil
Faculty of Economics and Management Sciences, University of Sfax, Sfax, Tunisia

Modils: Laboratory of Modeling and Optimization for Decisional Industrial and Logistic Systems, Sfax, Tunisia

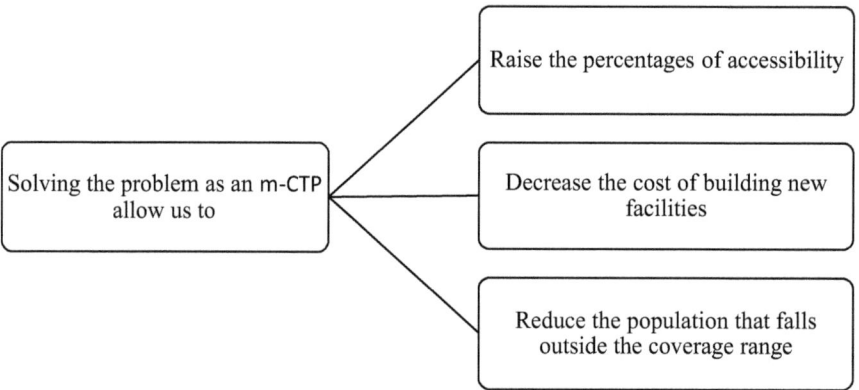

**Fig. 1** The advantages of our proposed solution

treatment of serious and disabling diseases requiring long-term care or particularly expensive care. These drugs can be found only in the specific pharmacy of the National Social Security Fund policlinic of Tunisia (NSSF). This pharmacy is responsible for distributing the treatments to the population of the 24 Tunisian cities. A large number of patients use to come to this pharmacy to take their treatment which causes congestion in the polyclinics.

Since 2007, the specific pharmacy for the specific drugs is created in the three main policlinics: El Omrane, Sousse and Sfax. Although there are three other policlinics, but they do not contain a specific pharmacy. Nowadays, considering the large number of the population, these three policlinics do not cover all regions and became not able to serve all patients.

In this paper, we are interested to solve this problem, and we propose to involve a set of pharmacies of the 24 Tunisian cities to distribute the treatments. More precisely, the demand of each city could be satisfied directly by visiting the city along the tour or indirectly by covering it. This constraint is one of the characteristics of the multi-vehicle covering tour problem (m-CTP). Our proposed solution has many advantages, which are presented in Fig. 1.

The remainder of this paper is organized as follows. In Sect. 2, a brief overview of the literature for the m-CTP is presented, after that, the model formulation of the problem. In Sect. 3, we first present our real case instances, then multiple tables are used to show the computational results. Finally, the paper is concluded and some future works are presented.

## 2   The Multi-Covering Tour Problem (M-CTP)

### 2.1   Overview of the Literature for the M-CTP

The CTP was first presented in Current (1981), where he proposed a heuristic method to solve this problem in which the constraint of visiting all the vertices is relaxed.

Few years later, Hachicha et al. (2000) proposed the formulation for the m-CTP using three-index vehicle flow formulation. After that, they presented three heuristics developed using different VRP techniques. Finally, to prove their efficacity, the different results of the heuristics are compared to each others.

Relying on column generation approach, Jozefowiez et al. (2007) proposed the first exact algorithm of the m-CTP. In this paper, the main problem was solved as a simple set covering problem while the subproblem was formulated basing on the CTP model of Gendreau et al. (1997).

Later, the m-CTP is studied differently by Hà et al. (2013). In this work, they considered only the restriction on the number of vertices per route, while the constraint of route's length is relaxed. In the first part, the authors used a branch-and-cut algorithm. In the second part, the authors proposed a metaheuristic considering the evolutionary local search (ELS) method to solve the problem. They proved the performance of their algorithm when solving different instances containing up to 200 vertices, while the tour could contain up to 100 vertices.

Murakami (2014) formulated the m-CTP as a set covering problem in the objective of saving time and life in each tour. In this work, the author developed a column generation algorithm in which the generated subproblem is solved by heuristics. Finally, he compared his results with those of the three heuristics of Hachicha et al. (2000), and they proved their performance in term of costs, unlike the computing time.

Another m-CTP variant, was introduced by Allahyari et al. (2015), which consist of the use of multiple depots. Two integer programming formulations were developed by the authors, using both flow and node. Also, a hybrid metaheuristic which combines GRASP, iterated local search and simulated annealing was generated.

Later, Kammoun et al. (2015, 2017) proposed a Variable Neighborhood Search (VNS) based on Variable Neighborhood Descent (VND) method, to solve the m-CTP-p, and the results proved to be better than those in Hà et al. (2013). More recently, the authors Kammoun et al. (2020) used a hybrid approach that combines GA and VND methods which they called (GA-VND) to solve the mm-CTP-p (multi-vehicle multi-covering Tour Problem with a constraint on the number of vertices). In the second part, they proposed a General Variable Neighborhood Search algorithm (GVNS) as a solution-based approach. The two meta-heuristics were compared to each other and with other approaches from the literature.

Flores-Garza et al. (2017) introduced the cumulative m-CTP, which aims to find a set of tours that must be followed by a fleet of vehicles in order to minimize the sum of arrival times at each visited location. The authors proposed a mixed integer linear programming formulation and a greedy randomized adaptive search procedure (GRASP).

Glize et al. (2020) proposed an effective exact method for the m-CTP based on column generation techniques. To validate their algorithm, they conducted extensive computational experiments on instances from literature. Also, the authors tried to solve the Bi-Objective Multi-vehicle Covering Tour Problem BO-m-CTP. In this second part, the objectives were minimizing both the total distance travelled and the maximum coverage distance.

Finally, in Margolis et al. (2021), a branch-and-price approach is developed to solve an m-CTP with speed optimization. The authors considered practical extensions to the model by incorporating risk thresholds, energy capacities, and time windows. Then, a two-stage heuristic was developed and it was capable of finding effective initial solutions.

## 2.2 Notation and Model Formulation

In our work, we deal with the m-CTP which is a generalization of the vehicle routing problem (VRP). The main objective of the m-CTP is to find a shortest route passing through a subset of vertices, subject to constraint on the number of vehicles used and the number of vertices including in these routes. The selected vertices are the subset which guaranties the maximum of coverage, because the non-visited vertex should be covered. From this we can understand that there are two subsets of nodes: some nodes must be visited by the vehicles, whereas some other nodes have to be covered. A node is covered if it is situated within a predefined covering distance from its nearest visited node.

The m-CTP is defined by Hà et al. (2013) as an undirected graph $G = (V \cup W, E)$, where $V \cup W$ is the vertex set and $E$ is the edge set. The vertex set $V$ is defined as $V = \{V_0 \ldots, V_{n-1}\}$ where $V_0$ is the depot, $m$ identical vehicles are located there, $\{V_1 \ldots, V_{n-1}\}$ are respectively the sets of policlinics and pharmacies (i.e., nodes to visit), and $W$ is the set of reminder cities (i.e., nodes to cover). Let $T \subseteq V$ is the set of policlinics that must be visited $T = \{V_1 \ldots, V_t\}$. For the set of edge, let $E = \{(V_i, V_j) V_i, V_j \in V \cup W, i < j\}$, for each edge of $E$ a length $(C_{ij})$ is associated. The problem consists of determining the shortest cycle by visiting some pharmacies from the set $V$ such that every city in $W$ is within a predefined distance $(d)$ from the cycle.

In such a problem a patient demand can be satisfied directly if he is in a city which is in the cycle, or from the pharmacy which respect the covering distance from his city. This distance can be defined as the largest distance that any person must travel to reach the nearest supplied pharmacy. In this work, the covering distance is determined such that each vertex of $W$ is covered by at least two vertices of $V\backslash T$. Then the problem is formulated by Hà et al. (2013) as follow:

$$\text{Minimize} \sum_{\{v_i,\,v_j\} \in \overline{E}} c_{ij} x_{ij} \qquad (1)$$

Subject to

$$\sum_{v_i \in V \setminus T} \lambda_{il}\, y_i \geq 1 \forall w_l \in W \qquad (2)$$

$$\sum_{v_i \in \overline{V}, i < k} x_{ik} + \sum_{v_i \in \overline{V}, j > k} x_{kj} = 2\, y_k \forall v_k \in V \setminus v_0 \qquad (3)$$

$$\sum_{v_j \in \overline{V}} (f_{ji} - f_{ij}) = 2 y_i \forall v_i \in V \setminus v_0 \qquad (4)$$

$$\sum_{v_j \in V \setminus v_0} f_{0j} = \sum_{v_i \in V \setminus v_0} y_i \qquad (5)$$

$$\sum_{j \in V \setminus V_0} f_{nj} = m * p \qquad (6)$$

$$f_{ij} + f_{ji} = p x_{ij} \forall \{v_i,\, v_j\} \in \overline{E} \qquad (7)$$

$$y_i = 1 \forall v_i \in T \setminus v_0 \qquad (8)$$

$$x_{ij} \in \begin{cases} 1 \text{ if edge } (v_i,\, v_j) \in \overline{E} \text{ is used in the solution} \\ 0 \text{ otherwise.} \end{cases} \qquad (9)$$

$$y_i \in \begin{cases} 1 \text{ if the vertex } v_i \in V \setminus v_0 \text{ is in the solution} \\ 0 \text{ otherwise.} \end{cases} \qquad (10)$$

$$fij \geq 0, fji \geq 0 \forall vi, vj \in E \qquad (11)$$

$$m \in \mathbb{N} \qquad (12)$$

The objective (1) of the m-CTP is to find a set of routes that minimize the total cost. Constraints (2) guarantee that every city of $W$ is covered. However, constraints (3) make sure that each pharmacy of $V$ is visited at most once. Concerning the flow variables is defined by the constraints (4–7). Constraints (4) confirm that at each pharmacy $\in V \setminus v_0$ the number of pharmacies already visited minus the number of pharmacies that can still be visited is equal to two if this pharmacy is visited and 0 otherwise. While constraints (5) define the number of pharmacies that can still be visited at the depot is equal to the total demand. Although the number of pharmacies already visited at the depot corresponds to the total capacity of the vehicle fleet confirmed by constraints (6). While constraints (7) confirm that the number of nodes on each route p in each pharmacy is equal to the summation of the pharmacies already visited and the pharmacies that can still be visited. Constraints (8) guarantee that each policlinic from the set *T* should be visited. Constraints (10–12) are used to Define the variables.

## 3 Computational Experiments

### 3.1 Preparing Instances

In this sub-section, we describe how we got our real case instances and after that the computational results. Our problem is solved as an m-CTP, but first of all some parameters have to be fixed before resolution. As cited in the Sect. 2.2, when talking about the m-CTP we must define three types of nodes *T*, *V*, and *W*. To help the decision maker, several scenarios are set to fix the sets of nodes.

- Two scenarios for *T*

    - Consider the used policlinics ($T = 3$)
    - Use all the available policlinics ($T = 6$)

- Three scenarios to determine the remainder sets *V* and *W*: Let *PP*, the percentage of patients in each node. We consider the nodes that their value of *PP* is less than 40%, 60%, and 80% to obtain the set *V*. The set *W* is remaining nodes.

    For each scenario, the number of nodes on each route is less than a value *p*. The parameter *p* is determined so that the routing cost cannot be improved ($p = 2,3,4...$).

    Our instances are identified *T–V–W–p*, in which *T* correspond to the number of policlinics used in the tour, *V* represents the number of the can be visited pharmacies, and *W* represents the number of the should be covered cities. Finally, *p* defines the number of cities on each route. For example, *T3–11–10–4* means that this instance requires 3 policlinics, 11 pharmacies that can be visited, 10 cities that must be covered and $p = 4$.

### 3.2 Results

We got 37 instances, which have been executed on a computer with an Intel Core i 7-2670QM, 2.20 GHz CPU and Ram 8 GB. But just 16 of them was successfully solved but it was not possible to solve the totality of the instances.

Both Tables 1 and 2 represents the computational results proposed for our real case, in which the column headings are as follows

- Data: Name of instances
- PP: the percentage of patients considered to obtain the set *V*.
- Coverage distance: calculated in kilometer
- Directly Served Patient"s Total Travel Cost

**Table 1**  The computational result for the first set of instances ($T = 3$)

| Data | PP | Coverage distance | Directly served patients | Total travel cost | CPU |
|---|---|---|---|---|---|
| T3–4–17–2 | 40% | 338 | 3,289,393 | 976 | 0.015 |
| T3–4–17–3 | | | | 692 | 0.01 |
| T3–4–17–4 | | | | 679 | 0.01 |
| T3–4–17–5 | | | | 679 | 0.005 |
| T3–11–10–2 | 60% | 211 | 3,478,488 | 1130 | 0.02 |
| T3–11–10–3 | | | | 909 | 0.02 |
| T3–11–10–4 | | | | 715 | 0.02 |
| T3–11–10–5 | | | | 702 | 0.025 |
| T3–11–10–6 | | | | 702 | 0.025 |
| T3–14–7–2 | 80% | – | – | – | – |
| T3–14–7–3 | | | | – | – |
| T3–14–7–4 | | | | – | – |
| T3–14–7–5 | | | | – | – |
| T3–14–7–6 | | | | – | – |

**Table 2**  The computational result for the second set of instances ($T = 6$)

| Data | PP | Coverage distance | Directly served Patients | Total travel cost | CPU |
|---|---|---|---|---|---|
| T6–4–15–2 | 40% | 303 | 4,236,776 | 1438 | 0.008 |
| T6–4–15–3 | | | | 1283.19 | 0.01 |
| T6–4–15–4 | | | | 1001.7 | 0.003 |
| T6–4–15–5 | | | | 991.7 | 0.004 |
| T6–4–15–6 | | | | 988.2 | 0.001 |
| T6–4–15–7 | | | | 987.2 | 0.001 |
| T6–4–15–8 | | | | 987.2 | 0.01 |
| T6–10–9–2 | 60% | – | – | – | – |
| T6–10–9–3 | | | | – | – |
| T6–10–9–4 | | | | – | – |
| T6–10–9–5 | | | | – | – |
| T6–10–9–6 | | | | – | – |
| T6–10–9–7 | | | | – | – |
| T6–10–9–8 | | | | – | – |
| T6–10–9–9 | | | | – | – |
| T6–13–6–2 | 80% | – | – | – | – |
| T6–13–6–3 | | | | – | – |
| T6–13–6–4 | | | | – | – |
| T6–13–6–5 | | | | – | – |
| T6–13–6–6 | | | | – | – |
| T6–13–6–7 | | | | – | – |
| T6–13–6–8 | | | | – | – |
| T6–13–6–9 | | | | – | – |

**Table 3** The computational results for the first set of instances ($T = 3$) using the second computer

| Data | PP | Coverage distance | Directly served Patients | Total travel cost | CPU |
|------|-----|------|------|------|------|
| T3–4–17–2 | 40% | 338 | 3,289,393 | 976 | 0.01 |
| T3–4–17–3 | | | | 692 | 0.005 |
| T3–4–17–4 | | | | 679 | 0.005 |
| T3–4–17–5 | | | | 679 | 0.001 |
| T3–11–10–2 | 60% | 211 | 3,478,488 | 1130 | 0.015 |
| T3–11–10–3 | | | | 909 | 0.015 |
| T3–11–10–4 | | | | 715 | 0.015 |
| T3–11–10–5 | | | | 702 | 0.02 |
| T3–11–10–6 | | | | 702 | 0.02 |
| T3–14–7–2 | 80% | 211 | 3,516,936 | 1001 | 0.025 |
| T3–14–7–3 | | | | 805 | 0.025 |
| T3–14–7–4 | | | | 694 | 0.03 |
| T3–14–7–5 | | | | 681 | 0.02 |
| T3–14–7–6 | | | | 681 | 0.025 |

- Total Travel Cost
- CPU: the computational time in seconds.

To solve all the instances, we had a problem with the Ram memory of our computer, for that we need another computer with a Ram memory more than 8 GB. for this we used another computer, Dell g3 gamer, with an Intel Core i7 and Ram 16 GB. Next, Table 3 and Table 4 shows the computational results of our instances obtained using the second computer, where all the instances were successfully solved.

By presenting many scenarios to the decision maker, we give him the possibility to choose the appropriate decision in each situation. For that we have to help him read these results, so we propose to compare three instances with the same value of $p$. For example, if we look at Table 3, it can be observed for these three instances ($T3$–4–17–5), ($T3$–11–10–5) and ($T3$–14–7–5) that the number of visited cities has increased from 4 to 11 to 14 this automatically indicates the growth of the number of directly served patients and the reduction of the coverage distance. In addition, we can notice that the number of cities that must be covered has decreased. However, if we compare the total travel cost for the same three instances, we notice that the cost has undergone a small variation. The same can be done for Table 4 for the second set of instances.

The increase of the total cost is insignificant comparing with the significant increase of the number of the directly served patients. It is clear that the third scenario is the best because it allows to cover the biggest number of patients with an acceptable total cost.

**Table 4** The computational result of the second set of instances ($T = 6$) using the second computer

| Data | PP | Coverage distance | Directly served patients | Total travel cost | CPU |
|---|---|---|---|---|---|
| T6–4–15–2 | 40% | 303 | 4,236,776 | 1438 | 0.008 |
| T6–4–15–3 | | | | 1283.19 | 0.01 |
| T6–4–15–4 | | | | 1001.7 | 0.003 |
| T6–4–15–5 | | | | 991.7 | 0.004 |
| T6–4–15–6 | | | | 988.2 | 0.001 |
| T6–4–15–7 | | | | 987.2 | 0.001 |
| T6–4–15–8 | | | | 987.2 | 0.01 |
| T6–10–9–2 | 60% | 211 | 4,425,871 | 1591.69 | 0.01 |
| T6–10–9–3 | | | | 1424.19 | 0.02 |
| T6–10–9–4 | | | | 1149.19 | 0.009 |
| T6–10–9–5 | | | | 1024.69 | 0.08 |
| T6–10–9–6 | | | | 1014.7 | 0.007 |
| T6–10–9–7 | | | | 1008.2 | 0.009 |
| T6–10–9–8 | | | | 1007.2 | 0.03 |
| T6–10–9–9 | | | | 1007.2 | 0.02 |
| T6–13–6–2 | 80% | 211 | 4,464,319 | 1553.5 | 0.01 |
| T6–13–6–3 | | | | 1298.69 | 0.01 |
| T6–13–6–4 | | | | 1114.5 | 0.01 |
| T6–13–6–5 | | | | 1003.7 | 0.01 |
| T6–13–6–6 | | | | 993.7 | 0.2 |
| T6–13–6–7 | | | | 987.5 | 0.023 |
| T6–13–6–8 | | | | 986.5 | 0.026 |
| T6–13–6–9 | | | | 986.5 | 0.029 |

## 4  Conclusions

In this paper, we have considered a real case study of the location of the specific pharmacy of the NSSF. This problem was solved as m-CTP using CPLEX solver. We started by preparing some different set of instances using multiple scenarios to help the decision makers of the NSSF to reach the appropriate solution according to the situation.

As a future work, we propose to implement a metaheuristic approach to avoid the problem of using a computer with very specific Ram configuration, which may not always be the case.

## References

Allahyari, S., Salari, M., & Vigo, D. (2015). A hybrid metaheuristic algorithm for the multi-depot covering tour vehicle routing problem. *European Journal of Operational Research, 242*(3), 756–768.

Current, J., (1981). *Multi-objective design of transportation networks*, PhD dissertation. Johns Hopkins University.

Flores-Garza, D. A., Salazar-Aguilar, M. A., Ngueveu, S. U., & Laporte, G. (2017). The multi-vehicle cumulative covering tour problem. *Annals of Operations Research, 258*, 761–780.

Gendreau, M., Laporte, G., & Semet, F. (1997). The covering tour problem. *Operations Research, 45*, 568–576.

Glize, E., Roberti, R., Jozefowiez, N., & Ngueveu, S. U. (2020). Exact methods for mono-objective and bi-objective multi-vehicle covering tour problems. *European Journal of Operational Research, 283*, 812–824.

Hà, M. H., Bostel, N., Langevin, A., & Rousseau, L. M. (2013). An exact algorithm and a metaheuristic for the multi-vehicle covering tour problem with a constraint on the number of vertices. *European Journal of Operational Research, 226*, 211–220.

Hachicha, M., John Hodgson, M., Laporte, G., & Semet, F. (2000). Heuristics for the multi-vehicle covering tour problem. *Computers and Operations Research, 27*, 29–42.

Jozefowiez, N., Semet, F., & Talbi, E. (2007). The bi-objective covering tour problem. *Computers and Operations Research, 34*, 1929–1942.

Kammoun, M., Derbel, H., & Jarboui, B. (2020). Two meta-heuristics for solving the multi-vehicle multi-covering tour problem with a constraint on the number of vertices. *Yugoslav Journal of Operations Research, 31*(3), 299–318.

Kammoun, M., Derbel, H., Ratli, M., & Jarboui, B. (2015). A variable neighborhood search for solving the multi-vehicle covering tour problem. *Electronic Notes in Discrete Mathematics, 47*, 285–292.

Kammoun, M., Derbel, H., Ratli, M., & Jarboui, B. (2017). An integration of mixed VND and VNS: The case of the multivehicle covering tour problem. *International Transactions in Operational Research, 24*, 663–679.

Margolis, J. T., Song, Y., & Mason, S. J. (2021). A multi-vehicle covering tour problem with speed optimization. *Networks, 79*(2), 119–142.

Murakami, K. (2014). A column generation approach for the multi-vehicle covering tour problem. *IEEE International Conference on Automation Science and Engineering*, 1063–1068.

# Applications of Transportation Models in Africa

Houda Alaya

**Abstract** Ample infrastructure, such as roads and proper vehicles to get goods to the right place at the right time, is an all-important factor in defining the quality of a transportation system. This has always been and continues to be a major challenge for developing countries such as most African countries. For decades, in Africa, urban development and transportation infrastructures have been inextricably interwoven. Cities began to grow, resulting in greater distances between activity locations and people's homes. The number of cars on the road and heavy traffic have increased even more, and transportation development projects are planned in different regions of Africa. Several papers have investigated the impact of African infrastructure on the transportation of materials such as oil and food and attempted to solve these problems by proposing mathematical models or algorithms that aimed to optimize one or more objectives, such as total travel cost or travel time.

In this chapter, we survey African transportation case studies and provide a critical perspective on the use of transportation models. We review real-life case studies in Africa related to transportation improvement and waste management. For each situation, we stress the recommended model or algorithm and then assess the outcomes of the proposed models and their impact on the African transportation progress.

## 1 Introduction

Various African countries have well-developed transportation networks, and these networks are organized to serve commercial and administrative interests.

H. Alaya (✉)
Business Analytics and Decision Making Laboratory, Tunis Business School, Tunis, Tunisia

© The Author(s), under exclusive license to Springer Nature Switzerland AG 2022   139
H. Masri (ed.), *Africa Case Studies in Operations Research*, Contributions to Management Science, https://doi.org/10.1007/978-3-031-17008-9_7

## 1.1  Africa Infrastructure

There exist several ways of transport in Africa (Hart, 2020).

- Animal Transport: Animals were and are still employed in various African countries. The usage of donkeys in Western and North Africa, camels in Western and North Africa, horses in Northern Nigeria, and bullocks in Southern Africa and the Horn of Africa have decreased as transportation infrastructure in African countries has improved.
- Motor Transport: The internal combustion engine's introduction and fast growth in the 1920s revolutionized the collection and distribution of products, as well as personal travel. Roads were constructed across North and Southern Africa, as well as areas of the west and east.
- Rail Transport: Early railways were built to help with the administration of interior regions and to transport goods from ports to central consumption or distribution centers, as well as to allow important minerals or commodities to reach the coast for sale, especially in the south. Rail networks, like roads, have developed significantly since the 1960s, resulting in decreased transportation costs.
- Air Transport: Air travel is perfectly adapted to Africa's enormous geography, and it has become the dominant mode of international and occasionally national travel throughout the continent. The fast growth of air transportation enhanced the flow of goods and people, as well as opened up the continent's hitherto isolated interior. Transportation became significantly faster and, in most cases, less expensive. Internal aviation services have continuously improved, while inter-continental air transport, particularly for people, has advanced significantly.
- Navigation: People and products were historically moved by canoe or boat on the various river systems throughout the vast interior between the Sahara and the Zambezi River. Engine-powered ships eventually supplemented or replaced canoes where conditions allowed, but water transportation has seen little further growth. Artificial harbors have been built on the shores in the meanwhile. A number of ports have been developed, and new berths have been added to existing port facilities.

Africa's poor infrastructure is one of the major roadblocks to its economic development. Moreover, Africa's transportation infrastructure is distributed inefficiently and unequally across the continent, with some parts over-equipped and others undeveloped (See Fig. 1).

Table 1 details the percentage of paved roads in different countries of Africa. We can observe that paved roads account for 13–33% of all roads in Africa. The variation in these percentages is due to many causes. Conditions notably include the geographical factors that are a major impediment to the construction of adequate highways and transportation systems.

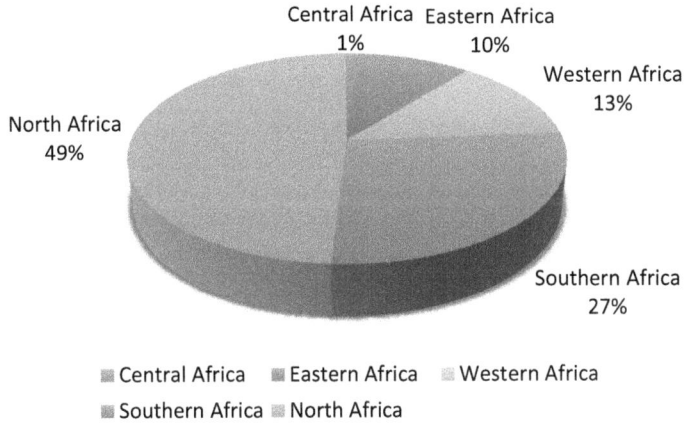

**Fig. 1** The percentage of paved roads in different sub-regions of Africa (Source: UN, 2009)

## 1.2 Impact of Africa Transportation Infrastructure on Economic Development

Africa is the primary recipient of international help, with transportation infrastructure projects accounting for the majority of international aid pledges to Africa. Foreign aid financing has been experimentally connected to beneficial economic outcomes in Africa. Table 2 is an overview of all the transportation projects in Africa undertaken by the World Bank.

The relationship between transportation infrastructure and economic development in Africa has been investigated in several papers. Ozment (2006) studied the link between economic growth and transportation in 44 emerging African nations. The GNP per capita was regressed on a set of factors that indicated changes in transportation from prior years. The relevance of the train network, paved motorways, and airports with fixed runways in explaining the current GDP/Capita demonstrates that transportation is critical for developing countries. Njoh (2008, 2009, 2000) investigated the relationship between "Sub-Saharan Africa's transportation infrastructure and economic growth in the era of globalization, the significance of Africa's transportation networks for development" and "In the context of Central and West Africa, development theory of transportation infrastructure," respectively. These studies, all of which focus on Africa, agree that a country's or region's transportation infrastructure has an impact on its development profile. They argue that improvements in transportation infrastructure, in particular, help to promote socioeconomic growth. In East Africa and Indian Ocean (EAIO) area, Njoh (2012) explored the link between transportation and development. It was anticipated that there is a positive relationship between developments, as measured by gross national income per capita (GNI/cap), and transportation, as measured by the several primary modes of transportation infrastructure in the research region. The hypothesis was

**Table 1** Percentage of paved roads in different countries of Africa (Data source: Central Intelligence Agency)

| Africa Country | 2014 | 2015 | 2016 | 2017 | 2018 | 2019 | 2020 |
|---|---|---|---|---|---|---|---|
| Central Africa Republic | | | | | 3% | | |
| Congo | | | | 13% | | | |
| Senegal | | | | 37% | | | |
| Eritrea | | | | | 10% | | |
| Malawi | | 26% | | | | | |
| Burkina Faso | 24% | | | | | | |
| Somalia | | | | | 0% | | |
| Burundi | | | 12% | | | | |
| Mauritania | | | | | 33% | | |
| Togo | | | | | | | 15% |
| Sierra Leone | | 9% | | | | | |
| Liberia | | | | | 6% | | |
| South Africa | | | 21% | | | | |
| Uganda | | | | | | 11% | |
| Kenya | | | | | 8% | | |
| Nigeria | | | | 31% | | | |
| Mali | | | | | 0% | | |
| Ethiopia | | | | | 0% | | |
| Zambia | | | | | 22% | | |
| Chad | | | | | 63% | | |
| Botswana | | | | 31% | | | |
| Madagascar | | | | | 0% | | |
| Mozambique | | 24% | | | | | |
| Sudan | | | | | | 26% | |
| Guinea Bissau | | | | | 10% | | |
| Eswatini | | | | | | | 0% |
| Equatorial Guinea | | | | 0% | | | |

tested using multiple regression with natural logarithms and associated statistics. At the 95% level, the resulting model was positive and statistically significant.

Several articles looked into the influence of African infrastructure on material transportation and sought to tackle these issues by presenting mathematical models or algorithms for real-world scenarios. In the following section, we will detail different transportation study cases in African countries.

## 2 Case Studies in Africa

Sustainable economic growth and poverty reduction address a complex set of challenges that necessitate the development of all sectors of the economy in order to fulfill the present generation's requirements without jeopardizing future

**Table 2** Transport projects list in Africa by World Bank (2022)

| Project Title | Project ID | Approval date | Last update date |
|---|---|---|---|
| Eastern Nile Flood Prevention and Early Warning Project —Phase 1 | P103518 | May 23, 2007 | March 10, 2016 |
| South Sudan–Eastern Africa Regional Transport, Trade and Development Facilitation Program (Phase One) | P131426 | May 20, 2014 | March 13, 2019 |
| NELSAP Water Resources Development Program | P116392 | February 4, 2010 | January 15, 2013 |
| Regional Trade Facilitation and Competitiveness DPO | P129282 | June 16, 2015 | December 27, 2018 |
| Strengthening ICPAC as a regional center of excellence for disaster risk reduction | P126851 | August 8, 2011 | March 19, 2014 |
| ENSAP (Eastern Nile Subsidiary Action Program) Facilitation | P069786 | October 12, 2006 | January 15, 2013 |
| ENSAP Multipurpose Program | P073460 | September 28, 2005 | January 15, 2013 |
| Tanzania-Transport Sector Support Project | P055120 | May 27, 2010 | January 18, 2018 |
| AFCC2/RI-East Africa Trade and Transport Facilitation Project | P079734 | January 24, 2006 | January 18, 2018 |
| Regional Development Project (03) | P000010 | February 1, 1990 | |
| Africa: CEMAC Transport and Transit Facil—Second Add'l Financing | P125915 | June 23, 2011 | January 18, 2018 |
| African EA&M Services | P100981 | June 27, 2005 | January 15, 2013 |
| 3A-ECOWAS Support for NEPAD Implementation (USAID TF) | P083066 | January 28, 2004 | January 18, 2018 |
| Urban Transport and Pollution Control | P076175 | | January 18, 2018 |

generation's long-term needs. One of the important sectors that play a critical role in poverty eradication and in reaching sustainable development goals is transportation. The transportation industry is intricately related to and impacts other economic sectors. In the literature, several researchers proposed solutions for real-world transportation services in order to reduce costs.

## 2.1  Routing Problems

Table 3 details solved real-life routing problems in different African countries.

**Table 3** An overview of the routing problems studied in Africa

| Authors | Country | Objectives | Proposed solution |
|---------|---------|-----------|-------------------|
| Groves et al. (2004) | South African | Minimize the total travel cost (time) | Graph-theoretic heuristic |
| Beukes et al. (2011) | Cape Town, South Africa | – | Geographic information system (GIS) based spatial multiple criteria evaluation (SMCE) |
| Masache et al. (2012) | Limpopo province of South Africa | Minimize the total cost | Sequential Insertion algorithm |
| Willemse and Joubert (2012) | Gauteng, South Africa | Minimize the total travel distance | Tabu search |
| Euchi and Mraihi (2012) | Tunisia | Minimize the total travel distance | Metaheuristic |
| Cao et al. (2016) | South Africa | Minimize the total cost | Algorithms |
| Masri et al. (2016) | Tunisia | Minimize the travel time | Recourse approach |
| | | Minimize the number of used vehicles | Chance-constrained approach Goal programming approach |
| | | Minimize the probability of accident | |
| Ben Abdelaziz et al. (2017) | Tunisia | Minimize the total travel cost | Recourse approach |
| | | Maximize the service quality | Goal programming approach |
| Catalani and Zamparelli (2017) | Mediterranean | Identify the best economically advantageous routing for oil product traders | Expert system |
| Semba and Mujuni (2019) | Tanzania | Minimize the travel time | Three Metaheuristics |
| Xin et al. (2021) | West Africa | Minimize the transportation cost | Active set algorithm |
| Zhu and Zhu (2022) | West Africa | Minimize the expected total cost | NSGA-II algorithm |
| | | Minimize the travel time | |

### 2.1.1 Vehicle Routing Problems

Over the last few decades, there have been a lot of studies of vehicle routing and scheduling issues. The Vehicle Routing Problem (VRP) is one of the first routing challenges. Initially, Dantzig and Ramser (1959) presented the VRP. The main goal is to determine the best route for a group of fuel delivery vehicles. VRP consists of providing a fleet of vehicles with various capacities, common depots, and various client requests. The goal is to develop the optimum vehicle route that starts at a common depot and distributes client demand while minimizing total trip costs. The road network for product delivery or pickup is represented by a graph with a set of vertices representing depots and consumers, as well as a set of connections. Several parameters relating to clients (demand, timeframe, and time for delivering and/or collecting product) and vehicles can be included in VRP models (home depot, capacity, compartments, and costs). The vehicle route should meet some constraints, such as capacity constraints (customer demands should not exceed the vehicle capacity during any route), duration constraints (the service time plus travel times should be less than a preset constant), and time window constraints (each customer should be served during a specified time interval). In the literature, there are various VRP variations which are, capacitated VRR; VRP with time window; VRP with Backhauls; VRP with pickup and delivery (See Fig. 2).

Groves et al. (2004) proposed a graph-theoretic heuristic for determining an effective service route for a single service vehicle via a transportation network with a subset of its edges that must be served a defined (possibly varied) number of times each. The periods at which each of these edges is to be serviced should be as equally spaced as is feasible over the scheduled time range, adding a scheduling element to the problem. The suggested heuristic was based on the Tabu search method, which is combined with a number of well-known graph-theoretic algorithms, such as Floyd's (for discovering shortest routes) and Frederickson's (for determining shortest paths). This heuristic was the foundation of a decision support system that asks the user for information about the physical situation (such as service frequencies and trip times for each network connection, as well as acceptable outcome boundaries), and then suggests a service routing schedule as an output. The decision support system was used in a special case study in which SPOORNET (the semi-private South African national railways authority and service provider) is looking for a service routing schedule for the South African national railway system as part of their rationalization effort in order to stay profitable. The Tabu search technique improved the first routing cost goal by 17% and improved also the initial temporal spread objective by 3.9%.

Masache et al. (2012) created an antiretroviral medicine delivery routing system with the purpose of increasing fleet utilization while lowering delivery costs. The method improved ARV medicine delivery satisfaction for patients in South Africa's Limpopo area. The problem was stated as a VRP mathematical programming problem, and the Savings Based and Sequential Insertion algorithms were employed to solve it. After that, a Visual Basic mini-program was created. Vehicle route determination heuristics were made faster using Internet software. In comparison to the pigeonhole method, which had a total traveled distance of 2874.2 km and a space usage of 86% for the needs of 5384 ARV medicine patients, this computer-based vehicle routing system had a total traveled distance of 1302.94 km and space utilization of 93%. As a result, the mathematical programming technique is more cost-effective and efficient, improving ARV medicine patient satisfaction in the province.

Willemse and Joubert (2012) presented a Tabu search algorithm capable of producing numerous patrol routes for an estate's security guards to address the absence of adequate patrol route planning methods. The difficulty of creating these

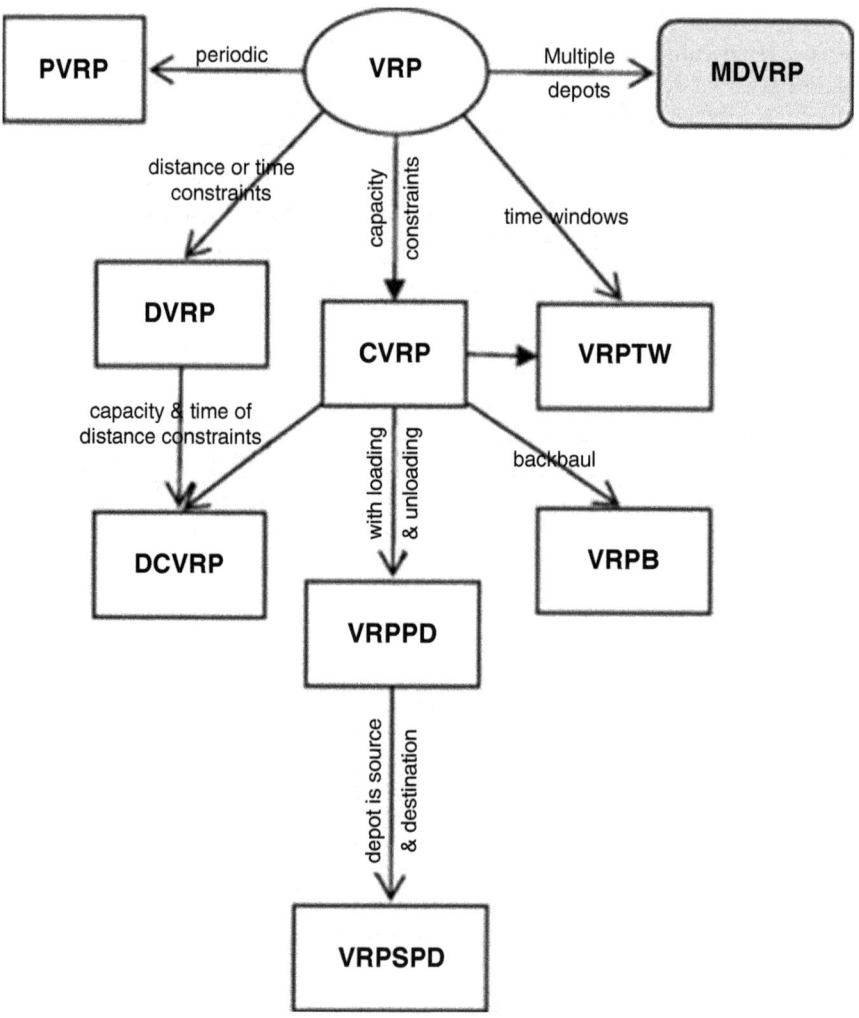

**Fig. 2** Vehicle Routing Problem variants (Source: Montoya-Torres et al., 2015)

routes may be modeled as an Arc Routing Problem, precisely as min-max k postmen problems. The technique was demonstrated using a real-world situation from a South African estate. The algorithm's patrol routes result in a considerable improvement in the even patrolling of the road network, as well as a more evenly distributed workload among guards.

The school bus routing issue (SBRP) in urban settings was the focus of Euchi and Mraihi (2012). Three simultaneous decisions have been made: identifying the set of locations to visit, determining which stop each student should walk to, and determining the routes traveled between the chosen stations, in order to minimize the overall traveled distance. As a result of the difficulties of solving the Tunisian case study, the authors used metaheuristic techniques, hybrid evolutionary computing based on an artificial ant colony with a variable neighborhood local search algorithm. The recommended solution regularly produced superior outcomes.

Cao et al. (2016) looked at how the supply chain for donated breast milk in South Africa may be expanded. The transportation of donated breast milk is hampered, as it is with the distribution of most public sector and humanitarian relief goods and services, by the inherent uncertainty in the environment, as well as the fact that, in addition to efficient resource use, the distribution must be made in an equitable manner. The scientists used many scenarios to include uncertainty in the models, which are based on various assumptions about population size, HIV prevalence, and the state of public health in the country, as well as income and education. To tackle the problem, several equity-based goals are evaluated, and rounding-based algorithms are provided. Cao et al. (2016) studied two delivery schemes: one that employs out-and-back transportation and the other that makes many stops along the route. They investigated the trade-offs between the objectives, as well as the consequences of alternative public health policies, network expansion budgets, and supply and demand assumptions, using computational experiments. Then they went over the teaching resources that came out of this work, which included a case study, a supply/demand calculation tool, and an interactive decision-making tool.

Masri et al. (2016) studied a multi-objective stochastic VRP in which many competing objectives are minimized at the same time, such as trip time, the number of cars in use, and the chance of an accident. Demands and travel times were considered stochastic variables. The authors proposed a solution technique based on a recourse approach, a chance-constrained approach, and a goal programming approach to developing a certainty comparable program to the multi-objective stochastic vehicle routing issue. The suggested MSVRP was created to address the needs of a Tunisian gas cylinder distributor. The Tunisian gas cylinder distribution company's operations entail shipping gas cylinders from a factory in Gabes to 13 distribution markets in Tunisia's various regions. The system delivered a high level of transportation safety and dependability, up to 97.3%, with just a 0.026 increase in the chance of accidents on specific routes.

The topic of airport bus routing (ABRP) was handled by Ben Abdelaziz et al. (2017). The following is a description of the ABRP: Customers are transported back to the airport by a firm that owns multiple buses stationed at the airport, according to their departure schedules, from a variety of hotels and meeting spots. Because the availability of clients at meeting places is unpredictable, the proposed problem may be understood as a stochastic vehicle routing problem. Its goal was to provide a low-cost set of vehicle routes that meets all customers' scheduling needs while reducing customer travel time and airport waiting time. The authors used a target programming technique and a recourse strategy to solve a multi-objective stochastic program (MSP) to simulate the ABRP. The suggested approach was put to the test using real-world data from a transportation business at the Tunis–Carthage airport. When compared to the prior solution employed by the firm, the model solution cut passenger travel and waiting times by 60%. As a consequence of the findings, the travel agency was able to better please travelers while lowering the number of vehicles used.

Simulated Annealing, Tabu Search, and Ant-Colony Optimization were three mega-heuristic algorithms utilized by Semba and Mujuni (2019) to address a real-world issue, the School Bus Routing Problem. Secondary data from Atlas Primary School, African Nursery and Primary School, and Yemeni Secondary School in Dares Salaam, Tanzania, were used to evaluate heuristics. The researchers looked at two scenarios: when the time to execute each heuristic is limited, and when it is not. Tabu Search performed badly when the time was limited but well when the time was not limited, according to the findings.

### 2.1.2   Other Routing Issues

Beukes et al. (2011) developed an approach for incorporating considerations such as the socioeconomic characteristics of the people served by the road and the environment in which the road is located into road design. Based on a mix of traffic and non-traffic characteristics, the technique aimed to prioritize the five key road-based modes (public transportation, vehicle, freight, pedestrian, and cycling). To arrive at modal priorities, the technique employed a spatial multiple criteria evaluation (SMCE) model based on a geographic information system (GIS) with inputs from generally accessible data sources such as census, household travel surveys, land use, and environmental data. An arterial highway in Cape Town, South Africa, is the subject of the case study. The investigation of the road in Cape Town reveals distinct variances in locational context along the route, as well as distinct portions of the route with distinct priority regimes. These distinctions highlight the significance of comprehending these contextual elements and the role they play in determining who uses the road and how it is utilized. Because there are so many contextual variables along the route, it is critical that these variances are assessed and included from the start when developing the amenities. According to the sensitivity analysis, the approach was capable of producing findings that reflect differences in weighting priorities without entirely distorting or drastically changing the unweighted outcomes.

Catalani and Zamparelli (2017) presented a system for oil product dealers to choose their most economically beneficial route. An expert system model was used in conjunction with the MNL (multinomial logit) technique. In the context of trade uncertainty, the expert system was used to address routing difficulties. The MNL model database, which includes leading shipping businesses in the oil derivatives market, is based on route selection using key characteristics such as distance, transportation cost, and terminal accessibility. The approach entails modeling routes provided by ships calling at MED (Mediterranean) ports in a discrete area.

Xin et al. (2021) proposed a model that optimizes the construction scheme of the regional inland transportation network (i.e., railway network), the expansion scheme of the port (i.e., container port) throughput capacity, and the operation scheme of the shipping network (i.e., container shipping network) given a limited budget. The authors built an algorithm to investigate the local optimum solution of the afore-mentioned model. Numerical tests were carried out to verify the efficacy of the model and method using China–West Africa transportation as the study target. For varied investment sizes and origin-destination demand, two types of sensitivity assessments were devised. The findings suggested that the proposed model and algorithm can successfully handle the aforementioned problem and can assist the Chinese government in developing suitable investment programs for West Africa.

Zhu and Zhu (2022) created multi-objective path decision models of multimodal transport under various carbon policies, taking into consideration the uncertainty in highway transport speed and transshipment time in the real transport process. The introduction of a low-carbon strategy has been shown to play a positive influence in lowering carbon emissions as a means of achieving sustainable transportation. However, a variety of factors influence the transportation of products, with unpredictably bad road conditions having a significant impact on both costs and time. To get the set of viable pathways and non-dominated solution sets of path schemes, the Path Search Algorithm and Fast NSGA-II were used. The efficiency of the suggested model and algorithms were evaluated using a case study of West African freight transit, and trade-offs were identified. The analysis of path schemes in non-dominant solutions revealed that improving the Pareto set was positively correlated with changes in highway speed; path schemes with carbon emissions exceeding the limit would be excluded under the implementation of a Carbon Cap policy, and only when the carbon tax rate was high enough could it influence the results of the path decision. The implementation of a Carbon Tax would significantly drive up the transportation costs. Carbon emission rules are being implemented in a way that encourages transportation companies to project low-carbon transportation plans and governments to establish related legislation. However, given the economic conditions in West Africa, the formation of a carbon-trading market was not feasible, and the implementation of a carbon tax policy may be difficult; consequently, it is more appropriate to adopt Carbon Cap and Carbon Tax policies. This paper assisted local governments in adopting reasonable carbon emission policies and forming a stable market order in combination with the current state of the local economy and transportation infrastructures by analyzing the impact of various carbon emission policies and uncertain road conditions on transportation routes.

## 2.2  Waste Management Problems

Waste Management contributes to minimizing pollution and promoting green logis-tics. It has consequently drawn the attention of the scientific community and has been extensively investigated over the past few years.

Green logistics can be defined as the design of products concerning the environ-ment, safety, and health requirements over the entire product life cycle, where we extend the well-defined supply chain management to include what is commonly called in the literature the reverse logistic. The waste management problem is part of the reverse logistic problem where we need to coordinate the flow of material and the flow of information through the nodes of a chain to collect, sort, pre-process, and dispose of the waste.

Usually, the waste is stored in waste containers or recycling bins and is collected using waste collection vehicles. These activities represent the waste collection step. The next step, the transportation step, begins when the collection vehicles leave the site and the last waste container has been unloaded. The duration of this transportation step corresponds to the time from the collection route to the disposal site. Unfortunately, waste collection and transportation operations are usually conducted without identifying the demand and specifying the transportation routes. These are usually left to the drivers of the waste collection vehicles.

The waste disposal step is divided into treatment and landfill. The treatment involves recycling, composting, and incineration. The recycling process consists of transforming the collected recoverable waste materials into reusable materials. In most cases, the recyclable waste, for example, paper, glass, plastic, and metal, can be reused, valorized, reconditioned, recovered, buried, repaired, or refurbished (Bereketli et al., 2011). Composting is a natural recycling process by which we transform the decomposed organic waste materials into a biological substance that can be used as a soil fertilizer. Composting is used for organic substrates such as municipal solid wastes, sewage sludge, and agricultural and industrial bio-products (De Meyer et al., 2015). Incineration is a waste treatment process that converts the waste into ash, flue, and heat, through the combustion of the waste materials (Vadenbo et al., 2014). Landfilling is the most common practice and most accessible solution to waste disposal (Muttiah et al., 1996). It consists of storing wastes under safe conditions for the environment and public health.

### 2.2.1   Urbanization's Impact on Waste Management

The increasing rate of unplanned and unregulated urbanization in Africa's emerging countries has resulted in environmental deterioration. Indeed, solid-, liquid-, and toxic-waste management has been one of the most serious challenges of urbanization in the developing world, particularly in Africa (Onibokun & Kumuyi, 1999). According to Gebremedhin et al. (2018) and based on the world bank report of 2012, the municipal solid waste (MSW) production in Africa was 125 million tons per year in 2012, with 81 million tons (65%) coming from Sub-Saharan Africa. Africa's waste production is expected to increase to 244 million tons per year by 2025 (See Fig. 2). In 2012, Africa's average MSW generation was 0.78 kg per capita per day, significantly lower than the global average of 1.2 kg per capita per day. However, due to variances in waste accounting, consumer attitude, economic level, and culture, there is a significant variety throughout the continent, ranging from 0.09 kg per capita per day to 3.01 kg per capita per day. By 2025, MSW output in Africa is expected to grow to 0.99 kg per inhabitant per day, a 1.27-fold increase over 2012. MSW in Africa (Sub-Saharan Africa) is roughly 57% organic, 9% paper/cardboard, 13% plastic, 4% glass, 4% metal, and 13% miscellaneous materials (See

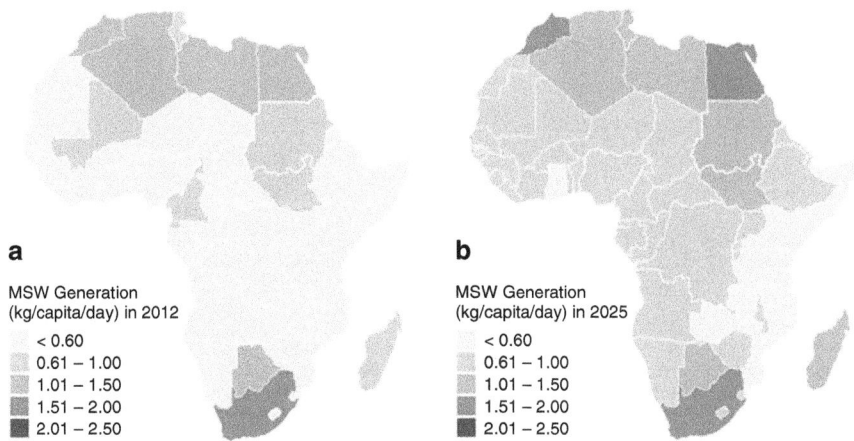

**Fig. 3** The spatial distribution of African nation's daily per capita garbage creation in 2012 (**a**) and 2025 (**b**) (Source: Gebremedhin et al., 2018)

Fig. 3). MSW in under-developed nations has a greater organic content than paper and packaging. However, the composition of MSW in Africa varies by location, based on consumer attitudes, economic levels, culture, and other factors. Despite the fact that per capita garbage creation in African cities is among the lowest in the world, demand for waste services continues to outstrip supply. Waste collection accounts for the majority of the money for solid rubbish management in poor nations, however, total waste collected in Africa (in 2012) was only 55% of total waste created (68 million tons). In Sub-Saharan Africa, the average MSW collection rate was 44%, while coverage varied greatly between cities, ranging from less than 20% to over 90%. In rural places, the situation is considerably worse. By 2025, the continent's average MSW collection rate is predicted to reach 69% (Fig. 4).

## 2.2.2 Solid Waste Management Problems

Waste management is a social, economic, and environmental issue that all African nations are grappling with. Weak organizational structures, a lack of sufficient skills, inadequate finances, and political instability are some of the current causes behind Africa's poor solid waste management (Godfrey et al., 2019). Several papers studied waste management in Africa. Nkosi et al. (2013), and FC et al. (2018) provided overviews of different waste management problems in Africa. Few papers solved solid real-life cases of waste management in African countries. Table 4 summarizes the studied problems.

Badrana and El-Haggar (2006) provided a model for a municipal solid waste management system in Port Said, Egypt. It incorporated the usage of the collecting station idea, which has yet to be implemented in Egypt. The suggested system was modeled using mixed integer programming, and its solution is achieved using MPL

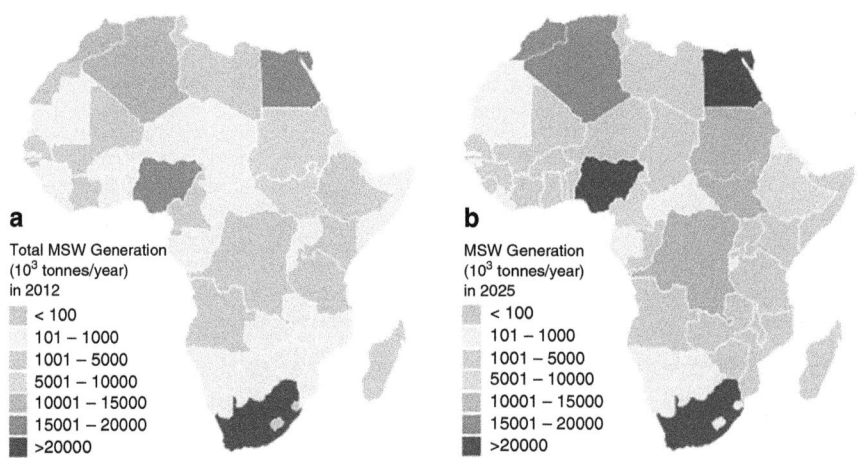

**Fig. 4** Africa's total MSW generation ($10^3$ tons/year) in 2012 (**a**) and 2025 (**b**) (Source: Gebremedhin et al., 2018)

software. According to the findings, the optimum model would comprise 27 collection stations with a daily capacity of 15 tons and two collecting stations with a daily capacity of 10 tons. There should be no trash transfer between the collecting point and the dump. Furthermore, the collection of district garbage should not be limited to district collection sites. For this approach, the goal function costs 10,122 LE per day (equal to US$1716).

Travers et al. (2008) focused on the optimization of waste management collection/transportation routing networks. They recommended using a geographic information system (GIS) 3D route modeling to optimize the route based on the least fuel consumption criterion to several settlements on the Cape Verdean island of Santo Antao. When compared to the shortest distance, the optimization for the lowest fuel consumption saves 52% of the fuel, despite traveling a 34% longer distance, demonstrating the importance of considering both the relief of the territory and the lowest fuel consumption criterion when optimizing vehicle routes. In this paper, fuel savings were significant with such a decision tool, management system efficiency was increased, and the environmental effect of everyday operations was decreased. The impacts of both the road incline and the vehicle load were taken into consideration by the GIS 3D route modeling.

Nemathaga et al. (2008) investigated the deficiencies in hospital solid waste management methods in the Limpopo Province of South Africa, using two hospitals as case studies. Aside from field surveys, the generated hospital trash was weighed to determine generation rates and tracked through various management strategies until final disposal. The findings indicated a significant policy implementation divide between the federal government and hospitals. While contemporary technologies such as landfills and incineration were employed, their everyday operations do not meet minimal requirements. Instead of being covered with soil, incinerator ash is

**Table 4** An overview of the Waste management problems studied in Africa

| Authors | Country | Characteristics | Model | Objectives | Objective function value |
|---|---|---|---|---|---|
| Badran and El-Haggar (2006) | Port Said Egypt | Subzones = 5<br>Collection Station = 29 (10 tons and 15 tons)<br>Containers = 12<br>Workers = 20 | Mixed integer programming | minimize the operational and fixed Cost of each facility. | Cost =10,122 LE/day (US$1716/day)<br>Profit = 49,655.8 LE/day (US$8418.23/day) |
| Tavares et al. (2008) | Cape Verde | Collection Station = 2<br>Vehicles Types = Trucks of 12 tons | 3D GIS model | Minimize fuel consumption | Fuel = 60,564 g per return trip<br>Significant savings in fuel of 9 per cent and 52 per cent |
| Nemathaga et al. (2008) | Limpopo Province of South Africa | 5 type of containers<br>3 Types of hospital waste<br>2 hospitals | – | – | General waste (60.74%) > medical waste (30.32%) > sharps (8.94%) |
| Tavares et al. (2009) | Cape Verde | Subzones = 2<br>Collection Station = 20<br>Vehicles Types = Trucks of 12 tons | 3D GIS model | Minimize fuel consumption | Fuel = 2686 g per trip<br>Praia City: Even though the distance traveled was 1.8% longer, the fuel savings was 8%.<br>Santiago Island: A 12% reduction in fuel consumption was achieved. |
| Jammeli et al. (2021) | Sousse Tunisia | Subzones = 13<br>Collection station = 738<br>Vehicles number = 13 (16 m$^3$) | Bi-objective stochastic model | Minimize the total cost and minimize the number of bins | Cost = 16001.573 TND per day<br>The strategy resulted in a 38% cost reduction. |

deposited openly, and wastes were burnt in landfills. The incinerators utilized were also unfriendly to the environment because they were outdated. Furthermore, the findings indicated that wastes were not properly separated according to their categorization, as required by the national government. The following was the average percentage composition of the garbage: general waste (60.74%) > medical waste (30.32%) > sharps (8.94%). The average generation rates per patient per day were determined to be 0.60 kg.

Tavares et al. (2009) recommended using GIS 3D route modeling software for waste collection and transportation, which provides another degree of flexibility to the system and allows driving routes to be optimized for the least amount of fuel use. The model took the impacts of road incline and vehicle weight into consideration. It was used in two separate situations: routing waste collection vehicles in Praia, Cape Verde's capital, and routing waste transport from several towns on Santiago Island to an incineration facility. The shortest 3D route, the 3D model that minimized fuel usage, resulted in an 8% cost savings in the Praia city region. Surprisingly, despite the fact that the GIS-recommended fuel-saving route was 1.8% longer than the lowest available trip distance, this was accurate. The difference was even more noticeable in the instance of Santiago Island: a 12% fuel savings for a similar overall journey distance. These studies showed how important it is to take into account both terrain relief and fuel usage when choosing a cost function for vehicle route optimization.

Jammeli et al. (2021) proposed a model for the problem of household waste collection and transportation in Sousse, Tunisia's largest city. The depot has a number of vehicles with limited capacity. Vehicles were required to collect rubbish that has been gathered in bins. After that, the waste is transported to a transfer station before the trucks return to the depot. The suggested model calculated vehicle routes and the number of bins to be assigned to each probable location, all while reducing collection costs and environmental effects. Because cost reduction was achieved by the optimal assignment of the established minimum number of bins, the problem was termed a bi-objective optimization problem. The stochastic component of population size, which is intended to follow a normal distribution, was investigated by Jammeli et al. (2021). Using a hierarchical strategy consisting of two steps as "cluster-first route-second," a solution was produced with an acceptable processing effort. The K-means clustering technique was used in the first step to divide a collection of n bin locations into k distinct groups. A certainly equivalent program to the bi-objective stochastic program was presented in the second stage. This was based on a chance-constrained, recourse, and goal programming technique. Using real data from the municipality of Sousse, the model was evaluated and applied. According to the findings, the strategy has a smaller environmental effect and reduces costs by over 38%.

Chikodzi et al. (2021) investigated the waste management systems in Durban, South Africa, and the responses of the various players, including the government, the business sector, and non-governmental organizations. Durban Harbor is one of Africa's most well-known and busiest ports, carrying freight for South Africa and numerous other southern African nations. Furthermore, its beaches are among the best in the world. Plastic contamination in the higher catchment areas is a visible indicator of inadequate waste management. Despite the fact that numerous parties have proposed several inventive remedies to the problem, they are mostly inadequate due to the problem's complexity. To secure the protection of the seas and the blue ocean economy, the municipality and stakeholders must step up their efforts to enforce environmental laws. The problem requires a multi-faceted strategy that includes, among other things, mass education, environmental law enforcement, and capital resource investment in coastal cleanups.

Chikodzi et al. (2021) recommended that while calculating the fuel consumption of garbage collection trucks, the terrain's variability should be taken into consideration. They considered that terrain elevation and route optimization through fuel consumption reduction are essential elements in the management of trash collection and transport vehicles in African countries. In Africa, waste management services and infrastructure should be carefully selected in terms of their long-term viability. More comprehensive, high-quality data on the volume, origins, kinds, and composition of trash created in Africa is needed, and this information should be shared across member nations.

### 2.2.3  Waste of Electrical and Electronic Equipment

Electronic waste (e-waste), also known as Waste of Electrical and Electronic Equipment (WEEE), is a major source of pollution in African countries, and the management of its environmental and health effects has been a concern. Electronic garbage is a worldwide issue, yet the harmful compounds it contains have local consequences. Because e-waste is still little-understood health and environmental issue, it receives less attention, particularly in African nations, making the youth more susceptible to dangerous substances. Bimir (2020) discussed the problem of electronic waste in African nations, the nature of present policy responses and constraints, and potential policy solutions. The study's data came from a variety of sources, including policy papers, law, and literature. The findings demonstrated that WEEE imports are increasing throughout Africa, but landfill and incineration remain the most common disposal methods. Nigeria and Ghana, two of the countries analyzed, lack WEEE-specific national regulations and adequate policy tools to ensure adequate collection and recycling. Despite the emergence of new recycling operations, a significant gap existed in that informal e-waste actors control the whole e-waste chain, from collection through to material extraction and refurbishment, using crude instruments that cannot identify harmful materials. To solve the problem, multi-actor interventions involving different stakeholders were needed to minimize WEEE influx on the one hand, while increasing safe recycling capacity on the other. As a result, the epistemic community was expected to write about it and generate knowledge in order to increase evidence for policy decision-making. Because the situation requires special attention, the focus is on Africa. E-waste has long been dumped in Africa, and those who work with it are often ignorant of, and exposed to, dangerous compounds. At the international level, the intellectual and policy communities must pay close attention to this issue.

## 3 Conclusion

This chapter discusses the impact of African countries' infrastructure on economic development and details the solution approaches developed by researchers to solve transportation problems, including routing problems and waste management problems. Regardless of all the efforts by world banking or by governments to ameliorate the transportation issues in African countries, they are still a challenge. Researchers are encouraged to present more and more research on this topic.

The healthcare industry is one of the most essential areas in achieving poverty eradication and sustainable development goals. The transportation industry is linked to the healthcare sector, and both influence other economic sectors. In our future research, we will be interested in the impact of healthcare criteria to develop economic development in African countries.

**Acknowledgments** We thank Donald Simpson of the Obex project for providing editorial support.

## References

Badran, M. F., & El-Haggar, S. M. (2006). Optimization of municipal solid waste management in Port Said–Egypt. *Waste Management, 26*(5), 534–545.

Ben Abdelaziz, F., Masri, H., & Alaya, H. (2017). A recourse goal programming approach for airport bus routing problem. *Annals of Operations Research, 251*(1), 383–396.

Bereketli, I., Genevois, M. E., Albayrak, Y. E., & Ozyol, M. (2011). WEEE treatment strategies' evaluation using fuzzy LINMAP method. *Expert Systems with Applications, 38*(1), 71–79.

Beukes, E. A., Vanderschuren, M. J. W. A., & Zuidgeest, M. H. P. (2011). Context sensitive multimodal road planning: A case study in Cape Town, South Africa. *Journal of Transport Geography, 19*(3), 452–460.

Bimir, M. N. (2020). Revisiting e-waste management practices in selected African countries. *Journal of the Air & Waste Management Association, 70*(7), 659–669.

Cao, W., Çelik, M., Ergun, Ö., Swann, J., & Viljoen, N. (2016). Challenges in service network expansion: An application in donated breastmilk banking in South Africa. *Socio-Economic Planning Sciences, 53*, 33–48.

Catalani, M., & Zamparelli, S. (2017). Maritime transport of oil and routing management. *Journal of Traffic and Transportation Engineering, 5*, 117–123.

Chikodzi, D., Dube, K., & Ngcobo, N. (2021). Rethinking harbours, beaches and urban estuaries waste management under climate-induced floods in South Africa. In *The Increasing Risk of Floods and Tornadoes in Southern Africa* (pp. 127–140). Springer, .

Dantzig, G. B., & Ramser, J. H. (1959). The truck dispatching problem. *Management Science, 6*(1), 80–91.

De Meyer, A., Cattrysse, D., & Van Orshoven, J. A. (2015). Generic mathematical model to optimise strategic and tactical decisions in biomass-based supply chains (OPTIMASS). *European Journal of Operational Research, 245*(1), 247–264.

Euchi, J., & Mraihi, R. (2012). The urban bus routing problem in the Tunisian case by the hybrid artificial ant colony algorithm. *Swarm and Evolutionary Computation, 2*, 15–24.

FC, O., Ogola, J. S., & Tshitangano, T. G. (2018). A review of medical waste management in South Africa. *Open Environmental Sciences, 10*(1).

Gebremedhin, K. G., Gebremedhin, F. G., Amin, M. M., and Godfrey, L. K. (2018). State of waste management in Africa. *United Nations Environment Programme.*

Godfrey, L., Ahmed, M. T., Gebremedhin, K. G., Katima, J. H., Oelofse, S., Osibanjo, O., & Yonli, A. H. (2019). Solid waste management in Africa: Governance failure or development opportunity. *Regional Development in Africa, 235.*

Groves, G., Le Roux, J., & Van Vuuren, J. H. (2004). Network service scheduling and routing. *International Transactions in Operational Research, 11*(6), 613–643.

Hart, J. (2020). Histories of transportation and mobility in Africa. In *Oxford research encyclopedia of African history.*

Jammeli, H., Argoubi, M., & Masri, H. (2021). A Bi-objective stochastic programming model for the household waste collection and transportation problem: Case of the city of Sousse. *Operational Research, 21*(3), 1613–1639.

Masache, A., Mashira, H., & Maposa, D. (2012). Antiretroviral drug distribution routing system in limpopo province of South Africa. *Journal of Mathematics and System Science, 2*(8), 512.

Masri, H., Ben Abdelaziz, F., & Alaya, H. (2016). A recourse stochastic goal programming approach for the multi-objective stochastic vehicle routing problem. *Journal of Multi-Criteria Decision Analysis, 23*(1–2), 3–14.

Montoya-Torres, J. R., Franco, J. L., Isaza, S. N., Jiménez, H. F., & Herazo-Padilla, N. (2015). A literature review on the vehicle routing problem with multiple depots. *Computers & Industrial Engineering, 79*, 115–129.

Muttiah, R. S., Engel, B. A., & Jones, D. D. (1996). Waste disposal site selection using GIS-based simulated annealing. *Computers and Geosciences, 22*(9), 1013–1017.

Nemathaga, F., Maringa, S., & Chimuka, L. (2008). Hospital solid waste management practices in Limpopo Province, South Africa: A case study of two hospitals. *Waste Management, 28*(7), 1236–1245.

Njoh, A. J. (2000). Transportation infrastructure and economic development in sub-Saharan Africa. *Public Works Manage. Policy, 4*(4), 286 296.

Njoh, A. J. (2008). Implications of Africa's transportation systems for development in the era of globalization. *Review of Black Political Economy, 35*(4), 147–162.

Njoh, A. J. (2009). The development theory of transportation infrastructure examined in the context of central and West Africa. *Review of Black Political Economy, 36*(3–4), 227–243.

Njoh, A. J. (2012). Impact of transportation infrastructure on development in East Africa and the Indian Ocean Region. *Journal of Urban Planning and Development, 138*(1), 1–9.

Nkosi, N., Muzenda, E., Zvimba, J., and Pilusa, J. (2013). *The current waste generation and management trends in South Africa: A review.*

Onibokun, A. G., & Kumuyi, A. J. (1999). Governance and waste management in Africa. In *Managing the monster: Urban waste and governance in Africa.* IDRC, .

Ozment, J. (2006). *Assessing transportation contributions to the economic performance of developing countries.* Final project report, college of business administration, University of Arkansas.

Semba, S., & Mujuni, E. (2019). An empirical performance comparison of meta-heuristic algorithms for school bus routing problem. *Tanzania Journal of Science, 45*(1), 81–92.

Tavares, G., Zsigraiova, Z., Semiao, V., & Carvalho, M. D. G. (2009). Optimisation of MSW collection routes for minimum fuel consumption using 3D GIS modeling. *Waste Management, 29*(3), 1176–1185.

Tavares, G., Zsigraiova, Z., Semiao, V., & da Graça Carvalho, M. (2008). A case study of fuel savings through optimisation of MSW transportation routes. *Management of Environmental Quality: An International Journal.*

Vadenbo, C., Guillén-Gosálbez, G., Saner, D., & Hellweg, S. (2014). Multi-objective optimization of waste and resource management in industrial networks–Part II: Model application to the treatment of sewage sludge. *Resources, Conservation and Recycling, 89*, 41–51.

Willemse, E. J., & Joubert, J. W. (2012). Applying min–max k postmen problems to the routing of security guards. *Journal of the Operational Research Society, 63*(2), 245–260.

Xin, X., Wang, X., Ma, L., Chen, K., & Ye, M. (2021). Shipping network design–infrastructure investment joint optimization model: A case study of West Africa. *Maritime Policy & Management, 1-27.*

Zhu, C., & Zhu, X. (2022). Multi-objective path-decision model of multimodal transport considering uncertain conditions and carbon emission policies. *Symmetry, 14*(2), 221.

# Operations Research Case Study Papers for Africa: A Bibliometric Review

**Majdi Argoubi and Hatem Masri**

**Abstract** Despite repeated requests for Operations Research to shift away from theoretical works and toward case studies, problem-oriented work, and real-world applications, theoretical papers still make up a sizable fraction of Operations Research publications. In order to accomplish two main goals: firstly, to investigate the scope of Operations Research case study in Africa (OR-CSA); and, secondly, to identify major areas of case study, evolutionary stages of the major techniques involved, and intellectual milestones in the development of key techniques. This work presents a systematic review of the literature on major aspects of OR-CSA. By using a generic search strategy, a representative dataset of OR-CSA bibliographic records is established. Next, we progressively synthesize empirical results. Results suggest that case studies are cited less often than other types of publication that deficit is more marked for African researchers where the case study has been effectively marginalized. It is suggested that, although a reduction in the proportion of publications in Operations Research made up of theoretical works may be desirable and would be an indication of maturity of the field, well-directed theoretical works will continue to play a role, albeit a diminishing one, in advancing the discipline. On the other hand, the evolution of the OR-CSA involves the development of several interconnected disciplines. As a final step, co-citation networks are constructed and visualized to assist with visual analysis of the OR-CSA's structural and dynamic relationships and developments. Fourteen major techniques are discussed in detail. For the purpose of demonstrating the analytical potential of the systematic method, the trajectory of citations made by specific categories of African authors and references is shown. Major milestones in key techniques are also investigated.

M. Argoubi (✉)
Higher Institute of Management of Sousse, University of Sousse, Sousse, Tunisia

H. Masri
College of Business Administration, University of Bahrain, Sakhir, Kingdom of Bahrain
e-mail: hmasri@uob.edu.bh

© The Author(s), under exclusive license to Springer Nature Switzerland AG 2022  159
H. Masri (ed.), *Africa Case Studies in Operations Research*, Contributions to Management Science, https://doi.org/10.1007/978-3-031-17008-9_8

# 1   Introduction

Operations Research (OR) was originally conceived during World War I (convoy theory and Lanchester's laws) to support military decisions (Rau, 2005). OR is literally the research of operations to improve decision-making and efficacy. There is a set of problem-solving techniques and methods, such as mathematical optimization, game theory, queuing theory, simulation, algorithmic solutions, stochastic-process models, neural networks, and other decision analyses, that improve the modeling of the operations and result in optimality and efficiency. Algorithms and the techniques that attempt to construct the mathematical models to describe the system and to solve the models are known as the optimization programming. Because of the stochastic and computational nature of most of these models, OR also has strong connections with computer science. Researchers confronted with a complex problem must determine which techniques are most suitable given the nature of the system, the objectives for optimization, and constraints on computing complexity and time. Or, they can develop a hybrid technique specific to the problem. The field application and interdisciplinary knowledge basis of OR are conducive to the modeling of complex systems and the optimization of decisions. Recent years have seen an increase in interest in OR from academics, businesses, and governments. At the same time, as computer science has developed, OR research has increasingly gained popularity as a hot area of study worldwide.

Subjective review and objective review are the two main categories used to classify prior OR review studies. Scholars' qualitative assessments of the essential strategies required to maximize aims, the range of applications, and the many tasks handled by the models have been variously incorporated in the research providing subjective reviews. However, there has not yet been a study that is specifically concerned with studying real-world applications, case study, and application areas. Furthermore, traditional literature review approaches only allow authors to comment on a limited set of studies, typically well within the domain of authors' experiences. Traditional literature reviews also suffer from subjectivity as they lack methodological rigor of a large-scale analysis. Subjective review can help overcome the limitations of a traditional literature review approach by utilizing an automated process that identifies case studies cohesions based on objective data.

The deployment of bibliometric databases and tools such as Web of Science (WoS) and Scopus, as well as the emergence of bibliometric analysis software, e.g., CiteSpace (Chen, 2004) and VOSviewer (Eck & Waltman, 2010), has started an era of extensive scientometric investigation analysis in all fields of research. This tendency has affected the OR field. Table 1 presents a wide variety of OR bibliometric surveys. Each bibliometric study can be focused on different objects, such as authors, institutes, countries, journals, and keywords. They can also focus on identifying relevant research topics and extracting the papers' content based on bibliometric data source, i.e. keywords, abstracts, or titles. The most frequent categories of scientometric analysis have been the study of the research output where the most prolific institutes, countries, and authors are identified and clusters

**Table 1** Previous bibliometric studies in OR field

| Reference | Scope | Aim | Unit of analysis | Data source |
|---|---|---|---|---|
| Liao et al. (2018) | OR field | Analysis of highly cited papers | Journals, countries, institutes, topics | Title, author keywords |
| Merigó and Yangj (2017) | OR field | Analysis of the field | Journals, authors, countries, institutes | Countries, journals |
| Shang et al. (2015) | OR field | Ranking of journals in OR | Journals | Countries |
| Kao (2008) | OR field | Analyze the most prolific countries and journals | Journals, countries | Countries, journals |
| Laengle et al. (2020) | OR field | Analysis of the most prolific universities | Institutes | universities |
| Romero-Silva and Marsillac (2019) | Journal in OR field | Evolution in the interest of OR field | Topics | Title, abstract, author keywords |
| Guan et al. (2019) | Journal in OR field | Clustering the core areas of research for journals | Topics | Abstract |
| Manikas et al. (2019) | Journal in OR field | Identify the most research areas in a journal | Research methods | Abstract |

of subjects are explored. Mainly with the help of bibliometric software, these studies have identified co-authorship relations among universities, the co-occurrence network of keywords, and co-citation networks among references. From bibliometric studies, we can see that the topics like mathematical programming, heuristics scheduling, simulation, routing, and decision-making have all been fundamental topics in OR. Findings from those analyses also indicate the emergence of themes such as sustainability, multicriteria decision-making, banking, and healthcare over the previous 10 years.

Argoubi et al. (2021) mention that "there is a lack of discussion on case studies in OR based on bibliometric analysis." This paper extends the results from previous bibliometric studies, which were focused only on most relevant research topics and output analysis, by studying the intellectual structure of case study in OR literature. Emphasis is placed mainly on African authors when dealing with OR case studies. Therefore, assuming no other study makes this claim, it is a unique endeavor to reveal the current state and of the real-world applications in OR-CSA research.

The remainder of this chapter is as follows: Sect. 2 provides data collection and the analysis methods; Sect. 3 presents the state of OR-CSA during the previous 32 years (1990–2022), including an examination of publishing and citation patterns, research trends at a disciplinary level, and the major contributors (co-authorship network, co-institute and co-country network); Sect. 4 shows research hotspots and research trend results, including co-citation analysis, timeline-view analysis, and the intellectual basis; Sect. 5 discusses the OR-CSA field's research frontiers and major milestones; finally, Sect. 6 summarizes key conclusions and proposes future expectations.

## 2 Data Collection and Analysis Methods

### 2.1 Data Sources

The approach to reviewing OR-CSA literature is based on bibliometric analysis. White and Mccain (1989) indicated that bibliometrics is a quantitative study of literature as captured in bibliographies. Various complementary advantages in using bibliometrics over a conventional review are also explained. Primarily, that research paper performance of institutions, individuals, and countries can be identified. Secondly, a broader and more mixed range of relevant topics can be investigated. Finally, a bibliometric overview allows a complementary point of reference.

Analyzing research works requires the extraction of the data from a research paper database. The input data for our review was created by combining the outcomes of several Web of Sciences (WoS) search queries. The WoS is a leading database and a source of high quality and multidisciplinary research information (Zupic & Cater, 2015). The logic of the query structure is as follows. We chose to combine one of the 54 African countries as "CU: Country/Region" with the search term "Operations Research" in order to extract OR research papers from Africa. This query generated 6658 records as Set #1. We note that the WoS database contains research works starting from 1990 and that the dataset was last updated in June 2022. The second query focused on the case studies in OR-CSA literature. By filtering out some records, such as program files, editing materials, reviews, and theoretical articles, a total of 5171 related articles were obtained. The research carefully scrutinized each article's title and abstract to ensure that the dataset it obtained was accurate and it found that all of the information was compliant. The final dataset was created from this database. This approach of query generation is robust enough to be used with OR-CSA. Other types of data sources that can be taken into consideration include research grants and patents. But, our review was limited to the scientific literature that the WoS has indexed. We postpone the evaluation of significance until the analysis step.

### 2.2 Methods

The study focuses on the bibliometric analysis of OR-CSA. This type of methodology has been used in previous studies. The methodology focuses on the documents published in the domain and analyzes various issues, including publication and citation trends, temporal evolution, authors, universities, and countries. In order to map the bibliographic material, we use the CiteSpace software. CiteSpace, a tool for measurement and visualization, uses a set of bibliographic records as input and models the intellectual structure as a synthesized network based on a time series of networks derived from each year's publications. This network is then used for the systematic evaluation of the knowledge structures. A broad variety of bibliometric indicators are used in this chapter to graphically display the research history, current developments, and hot themes in the field of OR-CSA. We created: an overview network, complete with a dual-map overlay and document co-citation analysis, for the systematic review of pertinent literature. The dual-map overlay displays academic activity across a number of disciplines in the subject under study (Chen & Leydesdorff, 2013). Co-citation (Small, 1973) appears when two documents receive a citation from the same third document. Co-citation clustering is automatically identified using spectral clustering methods based on spectral graph theory (Luxburg, 2007).

## 3  The Current Status of OR Case Studies in Africa

### 3.1  Publication and Citation Pattern Analysis

This analysis shows the growth pattern of publications over the years reflecting academic progress in the OR-CSA research field. As noted above, we found that a total of 5171 articles had accumulated over the past three decades (Fig. 1). There are three different file types that make up all the documents of OR-CSA research: articles, proceedings papers, and book chapters.

Publications on OR-CSA increased persistently over the years ($R = 0.9883$; $p < 0.05$). The change in the number of articles over time gives us a glimpse of

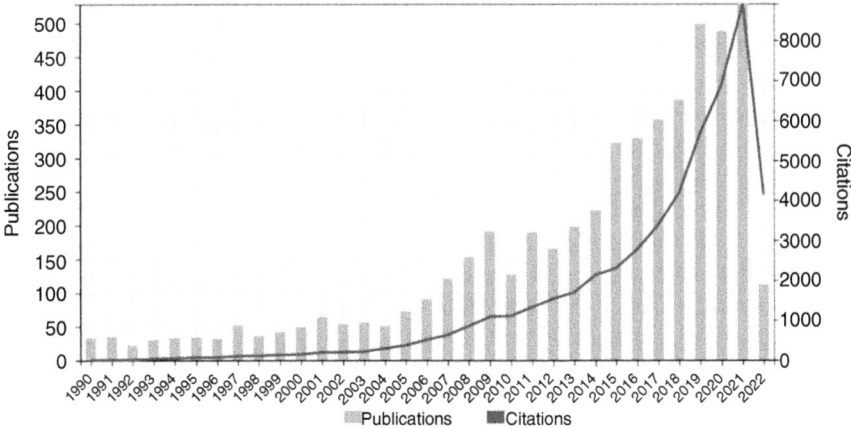

**Fig. 1** Times cited and publications over time

the interest and development of OR-CSA research. In total, the 5171 articles are cited 48,960 times apart from the self-citations until 31 June 2022, an average 9.89 times per article. The number of articles is expected to reach 6000 in 2023, because the average rate of increase reached 1 on average in each year from 1990 to 2022. Note further that our database, as large as it is, is limited to articles published in WoS. Thus, it does not represent all published knowledge on OR-CSA, but one can conclude that our field has accumulated a quite substantial body of knowledge.

We split the publishing rate into three periods based on the increased rate of papers published.

1. The embryonic phase (1990–2000). The publication activity slowly increased with relatively slow growth (no more than 50 papers within each year). At this stage, there is little attention on problem-orientated work and real-world applications, indicating that this research focus was only at its frontier development stage.
2. Steady development phase (2000–2010). The annual number of published articles increased steadily after 2000, with an average growth rate of 7%. During this period, a total of 1030 articles were published accounting for 20% of all the outputs (1990–2022).
3. Rapid development phase (2010–2022). The 528 publications in 2021 were more than 16 times those in 1990, showing remarkable growth.

These results indicate that research in OR-CSA, as an emerging domain, is receiving increasing attention and, consequently, more research is being conducted. However, compared to other research areas, research is still nascent. It is critical and urgent to make a good summary of the current status and analyze the future development directions in OR-CSA research.

## 3.2 Research Trends at a Disciplinary Level

A worldwide visualizing map of citation patterns at a domain level is shown by journal dual-map overlay analysis, which also illustrates how the citation distribution has changed over time (Chen & Leydesdorff, 2013). Through their citation relationships, the dual-map overlay reveals the articles' discipline concentrations. The global base map shows how more than 10,000 scientific publications are connected. These journals are further divided into areas that correspond to domain-level publishing and citation activity. A dual-map overlay allows us to see how several domains are related to one another. The frequency of citation scaled by the z-score affects a link's thickness (Kim et al., 2000). In this study, we analyze the knowledge diffusion patterns in the OR-CSA field over the years 1990–2022, from a global perspective using dual-map representations, as shown in Fig. 2.

The academic efforts of many disciplines in the OR-CSA field are shown in this picture. Individual journal themes are clustered closely together and each journal is represented in the diagram by a node. The citing component maps on the left and the

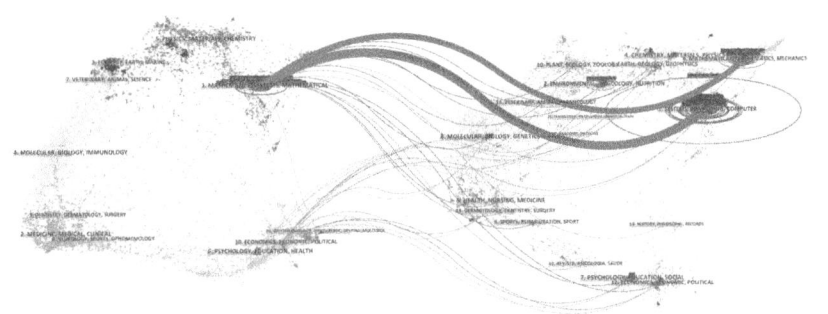

**Fig. 2** Dual-map overlay of citing literature and cited literature

**Table 2** Citation trends at a domain level

| Citing region | Cited region | Z-score |
|---|---|---|
| Mathematics/systems/mathematical | Systems/computing/computer | 5.20 |
| Mathematics/systems/mathematical | Mathematical/mathematics/mechanics | 2.17 |
| Economics/economic/political | All discipline | <0.5 |
| Psychology/education/health | All discipline | <0.5 |

cited component maps on the right are indicated by colored arcs that show the pathways of references. Because both citing and cited papers are classified into various subject categories, and because each position relates to one of them, the positions of these arcs illustrate how research is built upon prior studies. Each area is identified by a group of related journals, and each area is denoted by the phrases that are used most frequently in the names of the underlying journals.

These pathways are succinctly summarized in Table 2 together with citing and cited region names. The associations are arranged in decreasing order by z-score. In accordance with Fig. 2 and Table 2, literature in the OR-CSA domain is primarily represented on the citing map in three broad areas: the area in red near the top with the label "mathematics/systems/mathematical"; the area in blue in the lower position with the label "economics/economic/political"; and, the area in light blue with the label "psychology/education/health." The three areas' citation arcs point to specific regions on the right-hand cited map. The literature in "systems/computing/computer and mathematical/mathematics/mechanics" science journals serves as a significant foundation for "mathematics/systems/mathematical," as seen by the red citation arcs that are separated into two main streams. However, the citation arcs in blue and light blue converge for all disciplines. This indicates that OR-CSA research is an inter-disciplinary and cross-domain composite frontier area. The increasing complexity of research issues requires the intersection between disciplines.

## 3.3 Major Contributors to OR-CSA: A Collaboration Analysis

To overcome scientific challenges, to achieve major innovations, and to create interdisciplinarity in research, collaboration between different actors is fundamental. Katz and Martin (1997) defined scientific research cooperation as the working together of researchers to achieve the common goal of producing new scientific knowledge. Based on the collected literature data, this study intends to identify those scientific researchers, institutions, and countries that deserve attention and collaboration. The analysis of bibliometric records converts authors, institutions, and countries into nodes in order to produce the associated cooperation network map. The tighter and more effective the collaboration the larger the nodes, the greater the number of publications and the thicker the connecting link is.

The history of the field is reflected in the color of the node (citation ring). The thickness of the color and the corresponding time are proportional. The color changes from cold to warm. The density of a collaborative relationship serves as an identifier for each one. Density is a measure of connectedness inside a network and is calculated as the ratio of the number of actual linkages to the total number of links in a particular network. A stronger networking relationship is indicated by a density value that is closer to 1, whereas a density value that is closer to 0 denotes a less cooperative relationship.

### 3.3.1 Co-authorship Network

The most prolific authors play a leading role in a research domain and are considered to be the core driving force of the research in their field. In order to identify them, we used Price's law. Price's well-known square root law states that half of the literature on a subject will be contributed by the square root of the total number of authors publishing in that area (Newman, 2003). Further analysis revealed that authors who had more than 20 articles published were the most prolific in the OR-CSA field. Additionally, the data showed that 120 authors have published more than 20 publications, totaling 706 papers (accounting for 14% of the total amount). According to Price, it is realistic to expect that the authors who are most productive will produce 50% of the publications on a given topic. A stable core author group has not yet emerged in the OR-CSA sector as demonstrated by the clear difference between 14% and 50%. The authors who produced 20 or more papers between 1990 and 2022 are described in Table 3. These authors can be identified as influential authors in OR-CSA research. Finkelstein M published 98 papers and therefore ranked first, accounting for 1895%, while 4 other people vied for top five places with more than 40 publications each. There is little difference in the number of papers published by the top 20 authors. The author collaboration network may look at the degree of cooperation and relationship between various authors. There are 1070 authors (nodes), as seen in the top figure of Figs. 3, and 710 collaborative interactions

**Table 3** High-yield authors of the OR-CSA field

| Authors | Record count | % of 5171 |
|---|---|---|
| Finkelstein M | 98 | 1.895 |
| Mbohwa C | 52 | 1.006 |
| Haouari M | 46 | 0.890 |
| Ali MM | 44 | 0.851 |
| Cha JH | 43 | 0.832 |
| Chabchoub H | 34 | 0.658 |
| Levitin G | 34 | 0.658 |
| Pretorius L | 33 | 0.638 |
| Chelbi A | 29 | 0.561 |
| Hmamed A | 27 | 0.522 |
| Jemai Z | 25 | 0.483 |
| Stewart TJ | 25 | 0.483 |
| Ben Abdelaziz F | 24 | 0.464 |
| Jarboui B | 23 | 0.445 |
| Pretorius JHC | 23 | 0.445 |
| Shehu Y | 23 | 0.445 |
| Yadavalli VSS | 22 | 0.425 |
| Telukdarie A | 21 | 0.406 |
| Engelbrecht AP | 20 | 0.387 |
| Gadhi N | 20 | 0.387 |
| Guo BZ | 20 | 0.387 |
| Loukil T | 20 | 0.387 |

(links). The network's density was 0.0012, showing that just 0.12% of the possible connections in the OR-CSA network had really materialized. This also shows that, in the OR-CSA field, reasonably solid and substantial collaborations between academic teams have not yet been established. However, the bottom portion of Fig. 3 shows two important cooperative groups with four or more cooperating authors. The first group includes Chabchoub H, Hachicha W, Jarboui B, Abed M, Ghedira K, Loukil T, Abdelaziz F, Masmoudi F, and 12 other authors, these authors are all from Tunisia. Evaluating the efficiency of OR techniques is one of their study foci. The second group consists of Mbohwa C, Pretorius J, Telukdarie A, and Pretorius L, who are from South Africa and Zimbabwe. The study of cutting-edge technologies, particularly their incorporation into OR methods, such as big data, artificial intelligence, and virtual reality, has piqued these authors' strong interest.

Another intriguing conclusion is that collaborative ties typically exist amongst colleagues, indicating that an Africa wide academic community concerned with case study in OR has not yet formed in the continent and the cooperation relationship between the African authors is not close. In addition, even if it exists, the cooperative frequency of scholars' groups is very weak and the time that they started collaboration is comparatively recent. As mentioned below, all these findings indicate that the scholars conducting OR-CSA are dispersed and they have weak academic links.

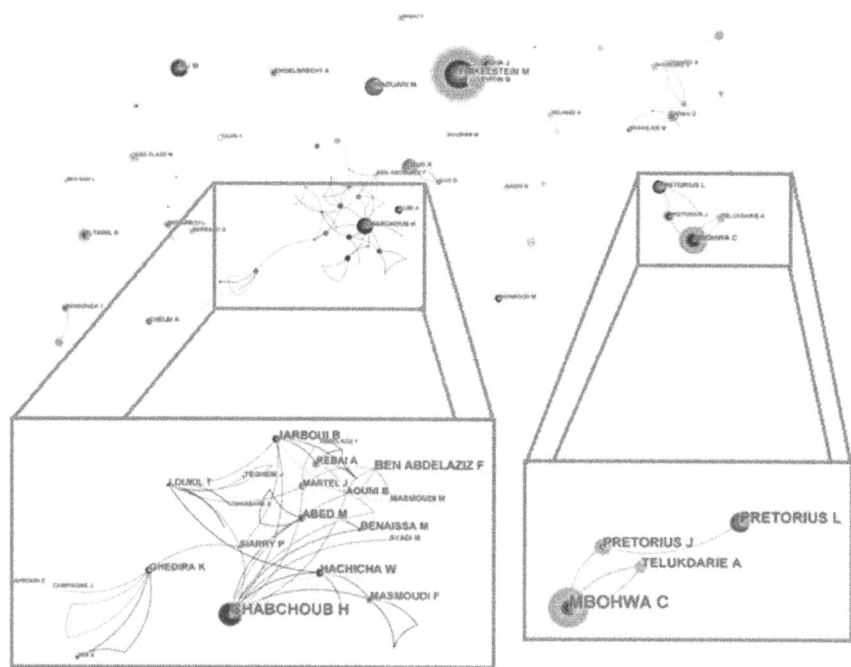

**Fig. 3** The author collaboration network and the zoomed top 2 major collaborative groups

**Table 4** The top 10 most
productive countries

| Countries | Record count | % of 5171 |
|-----------|--------------|-----------|
| South Africa | 1308 | 25.295 |
| Tunisia | 1039 | 20.093 |
| Morocco | 795 | 15.374 |
| Egypt | 775 | 14.987 |
| Algeria | 660 | 12.763 |
| Nigeria | 281 | 5.434 |
| Ghana | 62 | 1.199 |
| Cameroon | 35 | 0.677 |
| Zimbabwe | 34 | 0.658 |
| Kenya | 33 | 0.638 |

### 3.3.2 Co-institute Co-Country Network

It is evident that the more an institution or a country publishes, the more contributions it will make in the specific research field. The results show that the most productive countries are at their prosperous stages of development, and the most productive institutions are mainly from those countries. The development differs from countries to countries. Table 4 shows the top 10 most productive countries in this research field. South Africa is the most productive country followed by Tunisia,

**Table 5** The top 10 most productive institution

| Affiliations | Record count | % of 5171 |
|---|---|---|
| Universite de Sfax | 328 | 6.343 |
| University of Pretoria | 306 | 5.918 |
| Universite de Tunis el Manar | 213 | 4.119 |
| Universite de Tunis | 210 | 4.061 |
| Universite de la Manouba | 207 | 4.003 |
| University of Johannesburg | 191 | 3.694 |
| Cairo University | 164 | 3.172 |
| Mohammed v University in Rabat | 162 | 3.133 |
| Universite de Carthage | 149 | 2.881 |
| Sidi Mohamed Ben Abdellah University of Fez | 147 | 2.843 |

Morocco, Egypt, Algeria, Nigeria, Ghana, Cameroun, Zimbabwe, and Kenya. Research teams in South Africa and Tunisia are strong enough to develop collaboration systems or to conduct research independently.

Generally, the number of outputs is related to the number of research institutions, availability of research funding, and the proportion of those that have OR-CSA focus. At the institutional level, many institutions around Africa are engaged in OR-CSA research, and the top 10 productive institutions are presented in Table 5. Among these 10 institutions, there are five institutions in Tunisia, South Africa, and Morocco have two each, and Egypt has one. Tunisia's University of Sfax is ranked first in terms of research outputs. The maximum number of publications is only 328 from 1990 to 2022. It is a relatively small number of publications for one institute, indicating that the institute's research on OR-CSA has not gone deep enough.

For the analysis of the characteristics of OR-CSA, CiteSpace was adopted to visualize information based on authors' institutions. Figure 4 is a collaborative map of institutions in OR-CSA research. The analysis shows that scientific collaborations have been sensibly increasing in recent years. However, the OR-CSA knowledge base still suffers from a severe imbalance. Most of the nodes are isolated points, indicating that almost all the research results have been completed by a single institution. Only a few organizations have had collaborative experiences and the intensity of the existing cross-country links is very weak and not frequent. These links may be collaborative attempts, but are not yet sufficiently stable. Another interesting finding is that collaboration among institutions generally occurred between different universities of the same country. First, the University of Sfax had cooperated with University of Manar and University of Tunis. Together, they concentrate on research pertaining to the application of Genetic Algorithm, Goal Programming and various heuristic methods in routing problems, Knapsack problems, transport problems, and allocation problems. University of Pretoria established a partnership with University of Cape Town, University of Free State, and University of Johannesburg. Their main research interests include Control Charts and Analytical Hierarchy Process.

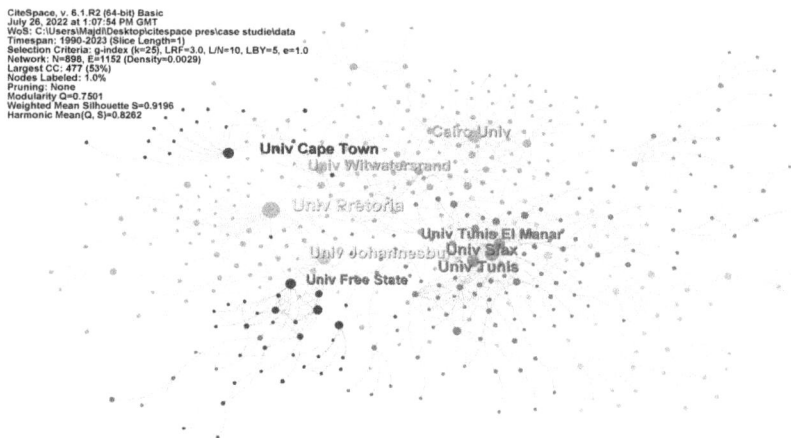

**Fig. 4** Top 2 major collaborative institutions

## 4 Research Hotspots and Research Trend: The Intellectual Base

### 4.1 Co-citation Analysis

Co-citation analysis consists of tracking couples of authors that are cited together in the source papers. When the same couples of authors are co-cited by several articles, groups of research begin to create. The co-cited authors in these clusters tend to contribute to the same idea (Chen, 2012). Combined with several clustering techniques, co-citation analysis can literally map the intellectual structure of specific research areas.

The following network is created based on papers published between 1990 and 2022 (Fig. 5) in the field of OR-CSA. The synthesized landscape view includes 1498 authors. To synthesize a network of authors cited in that year, top 100 most cited authors in each year are used. The colored arcs in Fig. 6 are co-citation connections that were added in the appropriate year as indicated by the arc's color. The thickness of an arc is proportional to the strength of co-citation. Large-sized nodes are of particular interest because they are highly cited.

The co-citation network includes 14 important clusters. The eight largest connected components contain 822 nodes which account for 55% of the entire network. Table 6 lists the top 14 most cited authors of the Co-citation clusters. Note that one may reach different insights into the nature of a co-citation cluster if different sources of information are used. The cited members of a cluster define its intellectual base, whereas citers to the cluster form a research front. Thus, Table 6 lists also the major citers of the important clusters. It supports analysts considering several aspects of the citation relationship from multiple perspectives.

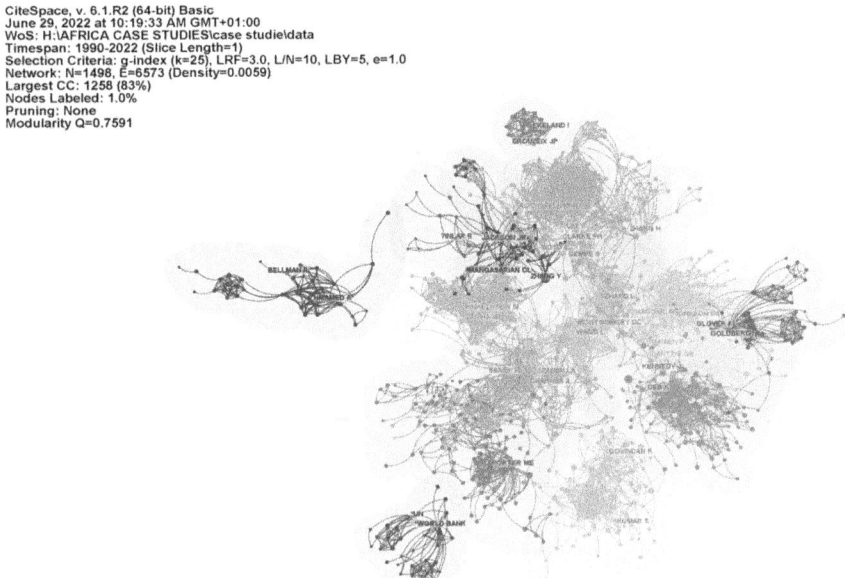

CiteSpace, v. 6.1.R2 (64-bit) Basic
June 29, 2022 at 10:19:33 AM GMT+01:00
WoS: H:\AFRICA CASE STUDIES\case studie\data
Timespan: 1990-2022 (Slice Length=1)
Selection Criteria: g-index (k=25), LRF=3.0, L/N=10, LBY=5, e=1.0
Network: N=1498, E=6573 (Density=0.0059)
Largest CC: 1258 (83%)
Nodes Labeled: 1.0%
Pruning: None
Modularity Q=0.7591

**Fig. 5** Author co-citation network

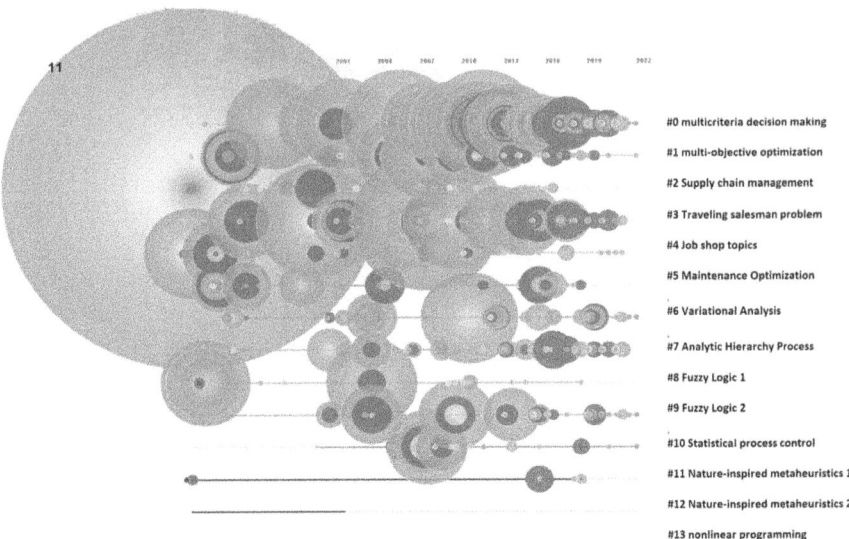

#0 multicriteria decision making

#1 multi-objective optimization

#2 Supply chain management

#3 Traveling salesman problem

#4 Job shop topics

#5 Maintenance Optimization

#6 Variational Analysis

#7 Analytic Hierarchy Process

#8 Fuzzy Logic 1

#9 Fuzzy Logic 2

#10 Statistical process control

#11 Nature-inspired metaheuristics 1

#12 Nature-inspired metaheuristics 2

#13 nonlinear programming

**Fig. 6** Timeline visualization of the network

The success of the clustering technique is evaluated by two metrics: (1) the modularity Q, which measures the extent to which a network can be divided into independent blocks (Newman, 2006). The value ranges from 0 to 1. A high value

**Table 6** Summary of the largest 14 clusters

| Cluster id | Size | Silhouette | Mean year | Most citing articles (Africa) | | Most cited authors | |
|---|---|---|---|---|---|---|---|
| | | | | References | Coverage (%) | Authors | Citation frequency |
| Cluster #0 | 127 | 0.897 | 2005 | Liu, D (2004) Object-oriented decision support system modeling for multicriteria decision-making in natural resource management. Computers & Operations Research | 14 | Porter ME (1999) | 35 |
| Cluster #1 | 120 | 0.867 | 2014 | Neggaz, N (2020) An efficient henry gas solubility optimization for feature selection. Expert Systems with Applications | 16 | Deb K (2007) | 99 |
| Cluster #2 | 105 | 0.883 | 2017 | Ruel, S (2021) Disaster readiness' influence on the impact of supply chain resilience and robustness on firms' financial performance: a covid-19 empirical investigation. International Journal of Production Research DOI | 30 | Govindan K (2016) | 21 |
| Cluster #3 | 102 | 0.815 | 2013 | Lahyani, R (2017) A unified matheuristic for solving multi-constrained traveling salesman problems with profits. Euro Journal on Computational Optimization | 10 | Dantzig GB (2006) | 47 |
| Cluster #4 | 96 | 0.854 | 2009 | Haouari, M (2006) Optimal scheduling of a two-stage hybrid flow shop. MATHEMATICAL METHODS OF OPERATIONS RESEARCH | 11 | Johnson DS (1993) | 46 |
| Cluster #5 | 92 | 0.931 | 2011 | Levitin, G (2020) Optimal mission abort policies for multistate systems. Reliability Engineering & System Safety | 14 | Finkelstein M (2011) | 88 |
| Cluster #6 | 92 | 0.898 | 2010 | Shehu, Y (2012) Iterative method for fixed point problem, variational inequality and generalized mixed equilibrium problems with applications. Journal of Global Optimization | 19 | Rockafellar RT (1990) | 90 |
| Cluster #7 | 88 | 0.897 | 2000 | Liu, D (2004) Object-oriented decision support system modeling for multicriteria decision-making in natural resource management. Computers & Operations Research | 13 | Saaty TL (1992) | 50 |
| Cluster #8 | 68 | 0.923 | 2009 | El, Azouzi R (2000) Perturbation of linear quadratic systems with jump parameters and hybrid controls. Mathematical Methods of Operations Research | 6 | Zhang H (2014) | 30 |

| Cluster #9 | 62 | 0.87 | 2010 | Bhrawy, A (2017) Numerical solution of the two-sided space-time fractional telegraph equation via Chebyshev tau approximation. Journal of Optimization Theory and Applications | 6 | Zadeh LA (2007) | 87 |
|---|---|---|---|---|---|---|---|
| Cluster #10 | 55 | 0.948 | 2015 | Aly, A (2015) A reevaluation of the adaptive exponentially weighted moving average control chart when parameters are estimated. Quality and Reliability Engineering International | 15 | Montgomery DC (2003) | 58 |
| Cluster #11 | 54 | 0.909 | 2010 | Mutingi, M (2013) A fuzzy particle swarm optimization approach for task assignment in home health care. 2013 IEEE International Conference on Industrial Engineering and Engineering Management | 11 | Clarke FH (2005) | 29 |
| Cluster #12 | 47 | 0.971 | 1999 | Sinclair, M (1993) Comparison of the performance of modern heuristics for combinatorial optimization on real data. Computers & Operations Research | 13 | Glover F (1993) | 48 |
| Cluster #13 | 43 | 0.943 | 1998 | Elbakry, A (1996) On the formulation and theory of the newton interior-point method for nonlinear programming. Journal of Optimization Theory and Applications | 9 | Zhang Y (1996) | 55 |

may indicate a well-structured network, whereas a low modularity indicates a network that cannot be reduced to clusters with well-defined boundaries (Chen, 2012). (2) The silhouette metric, which indicates the homogeneity of the network (Kaufman & Rousseeuw, 1990). The value of 1 represents a perfect separation between clusters, and therefore their themes (Chen, 2012). Each sub-domain in OR-CSA is well-defined in terms of co-citation clusters, as evidenced by our network's very high modularity of 0.75. Because there are so many little clusters, the average silhouette metric is rather low. However, the main clusters on which we will concentrate have a high enough silhouette score. Each cluster can be manually labeled based on common themes among its members. The areas are highlighted in various colors denoting the moments when co-citations initially occur. Areas in dark color were generated earlier than the light ones.

## 4.2   Timeline-View Analysis

A Timeline-View displays clusters along horizontal timelines (Fig. 6). Clusters are shown in a left to right direction. On top of the screen, the publishing time is displayed. With the largest cluster at the top, the clusters are placed vertically in declining order of their size (number of members). Visually, these clusters seem to be more interpretable than in the Co-citation network. Different domains have different levels of sustainability, as seen in the timeline-view. While some clusters last for more than 20 years, others are rather more transient. Even after the most recent year of publication in our analysis (2022), some domains are still in use. We will primarily concentrate on the five largest clusters in the next section. We will examine characteristics of each cluster such as the leading authors of a cluster as the intellectual basis.

## 4.3   The Intellectual Base

Latent semantic themes and overlaps can be discovered via cluster analysis (Hossain et al., 2011). There are a specific number of observed terms in each cluster (Van Waltman et al., 2010). The 14 largest clusters (Table 5)—which were regarded as the hottest topics over the 32-year study period—receive special attention in the discussion that follows.

Cluster #0 labeled "multicriteria decision-making" is the first largest cluster having a size of 127. The silhouette value is 0.897 and the mean year of this cluster is 2005. The active duration of the cluster starts from year 2010 and lasts up to 2020 (Fig. 6). This cluster contains numerous nodes with cold colors signifying that the research area is a recent research hotspot in this discipline's development. This cluster includes the following top research topics: multicriteria decision, multicriteria decision aid, decision analysis, human behavior, data collection, problem structuring, decision problems, maintenance management, decision support systems, and classification schemes. Furthermore, the most cited work in this cluster is by Porter (1990). The work of Michael Porter has towered over the field of competitive strategy. Also, it provides the foundation of the multicriteria decision-making analysis with the use of decision-making software. Our understanding of the dynamics of the each technique is aided by the articles that cited cluster members. The top citing article according to the cluster's bibliographic overlap, Liu and Stewart (2004), is an OR case study that discusses the object-oriented modeling of decision support systems for multicriteria decision-making in natural resource management.

The second-largest cluster, Cluster #1, labeled "multi-objective optimization" has 120 references spread across a 29-year span from 1993 to 2022 (Fig. 6). References in this cluster have 2014 as their median year. With a silhouette value of 0.867, this cluster is thought to have a high degree of homogeneity. The large and active cluster's main area of interest right now is the application of multi-objective optimization. The cluster's key concepts can be manually arranged using the titles of citing articles as a guide. The fundamental focus of this technique, for instance, is highlighted by ideas like conflicting objectives, Pareto-optimal solutions, and evolutionary optimization. The most cited article in this cluster is by Shukla and Deb (2007). The concept of optimization in the presence of multiple conflicting objectives is examined in this study and a single multi-objective optimization technique based on evolutionary algorithms is then presented. There are case studies offered for engineering design optimization. The breadth of case studies and the demonstrated discovery explain the high citation of this paper by researchers and practitioners who study similar engineering design problems.

"Supply chain management," Cluster #2, is the third largest cluster, with 105 references over a 26-year period from 1990 to 2016. When viewed in light of Fig. 6, the cluster or its underlying techniques are substantially active. Representative terms like logistics supply chain, supply chain systems, green supply, competitive advantage, robust optimization, genetic algorithms, and performance measures predominate in this cluster. The most cited article in this cluster is by Govindan (2016). This work investigates current trends in utilizing the evolutionary algorithms in supply chain management problems including complex networks and efficient algorithms.

The fourth-largest cluster, Cluster #3, titled "Traveling salesman problem" spans a 26-year period from 1995 to 2022 and has a size of 102. This cluster's median year is 2013. Despite being the lowest of the major clusters, the silhouette value of 0.815 is nonetheless regarded as having a pretty good level of homogeneity. This cluster includes the following top research topics: linear programming heuristic algorithms, genetic algorithms, exact algorithms, empirical evidence, comparative analysis, integer linear and survey classification. The most cited work in this cluster is by Dantzig et al. (1954). The work on the traveling salesman problem was the precursor of the branch-and-cut algorithms that form the basis of modern mixed-integer computational system. Those are widely used in practice to solve optimization models in telecommunications, manufacturing, transportation, and many other case studies.

Cluster #4 labeled "Job shop topics" is the fifth-largest cluster having 96 citing references. The mean year and silhouette score of this cluster are 2009 and 0.854, respectively. This cluster starts from year 1990 and lasts to 2020. This cluster includes the following main research topics: flow-shop scheduling, flexible manufacturing, resource constraints, setup times, supply chain systems, survey classification, classification scheme, project scheduling, linear programming, and material handling. The most cited article in this cluster is by Johnson (1993). This article proposes a computer-aided approach to the analysis of higher-dimensional domains, using several open problems about the average-case behavior of the Best Fit bin-packing algorithm as case studies.

With 92 cited references and a silhouette value of 0.931, Cluster #5, titled "Maintenance Optimization" is the sixth largest cluster and has a lower homogeneity than the five previous larger clusters. The mean year of this cluster is 2011. The majority of top citing references of this cluster are mathematical methods in reliability, risk and safety modeling, survival analysis, and stochastic processes. The most cited work in this cluster is by Finkelstein (2011). The concept of minimum repair is generalized in this work to the situation in which the lifespan distribution function is a continuous or discrete combination of distributions. It takes into account the statistically minimal repair and the minimal repair based on data obtained immediately prior to an object failing. Few straightforward examples are considered.

The other clusters are either modest in size or have a brief lifespan. We shall not go into depth about these clusters. Following are a few of the more pertinent clusters. Cluster #6 focuses on "Variational Analysis" as inspired by problems of optimization, with emphasis on applications to stochastic programming, optimal control, finance, economics, and engineering. The papers that are most frequently mentioned in this cluster were largely published in the early 1990s.

Cluster #7 focuses on "Analytic Hierarchy Process," led by the book of Saaty (1988). 2013–2020 saw the peak of activity for this cluster. Saaty TL, Charnes A, and Geoffrion AM, who are among the top contributors of this technique, are still actively involved in the cluster, particularly in operations management, robust optimization, case study, network analysis, and process design sub-problems.

Studies on "Fuzzy Logic" are included in clusters #8 and #9. These hotspots focused on mathematical programming, robust optimization, machine learning, particle swarm, mass customization, deep learning, and support vector. Research on Statistical process control is included in cluster #10 and this hotspot lasted the longest. Main keywords included control charts, statistical process control, process control, quality management, quality control, design process, and manufacturing processes. Studies on "Nature-inspired metaheuristics" are included in Clusters #11 and #12. The popularity of these techniques increased in the past 10 years. It focused on genetic algorithms, ant colony optimization, particle swarm, Monte Carlo, neural networks, engineering applications, and discrete optimization.

Finally, Cluster #13 contains 43 references with a median year of 1998; it is the cluster that contains the oldest articles. This cluster focuses on studies of nonlinear programming.

# 5    Research Frontiers and Major Milestones

OR is a highly dynamic research field. New cases studies continually emerge; others gain or fail in importance, split or fuse. It is challenging to keep track of the dynamic evolution and structure of OR-CSA since the number of publications is always rising. But an important factor in making decisions is being aware of hot topics, new research frontiers, or shifting focus. Our analysis applies a temporal metric to decipher co-citation networks and consequently generated clusters. The temporal metric reflects citation burstiness. The burst metric detects whether and when the citation count of a reference or a particular connection has surged significantly during a short time interval within the overall time period (Kumar et al., 2006). The burst detection algorithm introduced in Kleinberg (2002) was adopted in our study. In Fig. 6, a red ring corresponds to the occurrence of a citation burst. The list of references with significant increases in citations between 1990 and 2022 may be used to identify significant major milestones in the development of OR-CSA (Table 7). References with high values in the Strength column are frequently important turning points in OR-CSA research. In the following, we will present three of them.

The first milestone paper is a landmark study in Swarm Algorithm and belongs to cluster #1 labeled "multi-objective optimization." The article by Mirjalili et al. (2017) had a very high burst score for 4 years, unsurprisingly in light of the subject. In order to solve optimization problems with single and multiple objectives, this paper introduces two innovative methods called Salp Swarm Algorithm (SSA) and Multi-Objective Salp Swarm Algorithm (MSSA). The paper considers solving several challenging and computationally expensive engineering design problems. The paper demonstrates the ability of the proposed algorithms to address real-world problems. At the time of writing, that study has received 2322 citations. It has been widely cited and constituted a technological breakthrough, describing a procedure that came into general use for many Multi-objective-based case studies. It comes as no surprise, then, that this early work has accumulated the highest citation scores over long periods and has remained highly influential long after publication. It

**Table 7**  Top-10 Burst References in OR-CSA

| Cited Authors | Strength | Begin | End | 1990–2022 |
|---|---|---|---|---|
| Mirjalili S | 13.68 | 2018 | 2022 | |
| Haouari M | 12.04 | 2005 | 2012 | |
| Barlow RE | 11.63 | 2001 | 2012 | |
| Zadeh LA | 11.44 | 2007 | 2011 | |
| Guo R | 10.73 | 2005 | 2010 | |
| Goldberg | 10.31 | 1999 | 2011 | |
| Yang L | 10.23 | 2020 | 2022 | |
| Liu B | 9.98 | 2007 | 2011 | |
| Laporte G | 9.95 | 2014 | 2018 | |
| Finkelstein MS | 9.78 | 1999 | 2009 | |

contains the background for further intellectual and technological developments in Multi-objective optimization.

The second highest burst paper, which appeared in 2005, is closely related to the Job shop topic. Published in the International Journal of Production Economics, Al-Fawzan and Haouari (2005) discuss the problem of creating a project schedule that is not just brief in time but also less susceptible to delays brought on by reworks and other unfavorable circumstances. This study established the idea of schedule robustness and constructed a bi-objective resource-constrained project scheduling model. The objectives of robustness maximization were considered along with makespan minimization. A tabu search algorithm was developed in order to generate an approximate set of efficient solutions. This article has received 341 citations at this writing, and unsurprisingly has a strength burst value. It also directly connected to two other bursty studies identified in the present analysis.

Finally, the third top-most reference by burst is Barlow and Proschan (1996). This reference belongs to cluster #5 labeled "Maintenance Optimization." The citation burst started in 2001 and ended in 2012 as shown in Table 7. As a pioneer in mathematics of reliability and a laureate of the John von Neumann Theory Prize, Richard E. Barlow has a high degree of burstness. His research focuses on reliability mathematics, which examines the likelihood that a system or component will function effectively across an interest-precise time period.

# 6 Conclusion

OR-CSA is a research area with a lot of interest and potential impact. As this study demonstrates, OR-CSA offers researchers a broad and engaging research environment. The histories of economy, science, and technology all indicate that any new technology research and its application has an overall impact on the real society only after going through several cycles of development. Therefore, it is expected that the effects of OR-CSA methods on the economy and society will take time to develop. It is crucial for researchers to be up to date with trends in order to, at any time, gain knowledge about the research directions. To the best of our knowledge, we are among the first in the OR-CSA community to include evolutionary processes into a visual and analytical research by combining numerous complementary bibliometric methodologies, such as literature citation analysis, co-citation analysis, and burst analysis. Due to its hybrid methodology, our study is more valid.

Accordingly, we highlight the contribution of our study for academic advancement. In order to assist scholars to easily comprehend the evolution of OR-CSA and support future academic innovation, we first sum up a panoramic view of OR-CSA. In reality, a lot of researchers are interested in learning more about OR-CSA and are eager to investigate new academic prospects, but they only have a general knowledge of OR-CSA. Therefore, investigating new topics and perspectives for research may be done much more quickly, thanks to the authors, literature hotspots, and trends discovered in our study, giving practitioners a competitive advantage. Finally, our list of key research institutions and researchers serves as a resource for organizations looking to collaborate with research institutions, particularly those that are eager to explore and conduct research.

According to our analysis in this study, there has been a notable rise in the OR-CSA knowledge domain literature, notably after 2010. But there hasn't been much scholarly collaboration, and the research has been dispersed. Many high-yield universities, like Pretoria University and Johannesburg University, are located in South Africa, which has made the largest contribution. Tunisia and Morocco have conducted more study on OR-CSA than other regions. Additionally, the OR-CSA discipline has produced some notable researchers, like M. Finkelstein, C. Mbohwa, and M. Haouari. Other recent research hotspots include Multi-Objective optimization, Supply chain management, Traveling Salesman Problem, Job Shop Topics, Maintenance Optimization, Variational Analysis, Analytic Hierarchy Process, Fuzzy Logic, Statistical Process Control, Nature-inspired metaheuristics, and Nonlinear programming. Even though many different works have contributed to the growth of OR-CSA research, specific subjects represent the research frontiers and major milestones of the domain. After using the burst metric screening process, these contributions were identified as belonging to the Multi-Objective optimization, Job shop, and Maintenance Optimization themes.

In conclusion, bibliometric analysis is very important for determining potential connections between the literature and examining the knowledge evolution of OR-CSA study. There are still certain restrictions with this research, it should be noted. The offered data sources and the utilized query methodologies limit the breadth of this study because the findings are based on a single dataset. The quality and quantity of the data can be increased in the future by researchers using more datasets and new query techniques. The results in this research would also benefit from a more in-depth analysis in a few years as the OR-CSA field is anticipated to grow rapidly. Finally, due to the limits of the Bibliometrics technique, the influence of recently published literature can only be detected over time, and certain newly published papers with great potential of influence cannot be detected at this phase. However, we expect the development of new bibliometric techniques to assess the potential of these papers soon.

**Acknowledgments** We thank Paul Randall of the ōbex project for providing language editing support.

# References

Al-Fawzan, M. A., & Haouari, M. (2005). A bi-objective model for robust resource constrained project scheduling. *International Journal of Production Economics, 96*(2), 175–187.

Argoubi, M., Ammari, E., & Masri, H. (2021). A scientometric analysis of Operations Research and Management Science research in Africa. *Operational Research: An International Journal, 21*, 1827–1843.

Barlow, R. E., & Proschan, F. (1996). *Mathematical theory of reliability*. Society for Industrial and Applied Mathematics.

Chen, C. (2004). Searching for intellectual turning points: Progressive knowledge domain visualization. *Proceedings of the National Academy of Sciences of the United States of America, 101*, 5303–5310.

Chen, C. (2012). *Turning points: The nature of creativity*. Springer-Verlag.

Chen, C., & Leydesdorff, L. (2013). Patterns of connections and movements in dual map overlays: A new method of publication portfolio analysis. *Journal of the American Society for Information Science and Technology, 65*(2), 334–351.

Dantzig, G. B., Fulkerson, D. R., & Johnson, S. M. (1954). *Solution of a large-scale traveling-salesman problem*. RAND Corporation.

Eck, N. J. V., & Waltman, L. (2010). Software survey: VOSviewer, a computer program for bibliometric mapping. *Scientometrics, 84*(2), 523–561.

Finkelstein, M. (2011). On the 'rate of aging' in heterogeneous populations. *Mathematical Biosciences, 232*(1), 20–23.

Govindan, K. (2016). Evolutionary algorithms for supply chain management. *Annals of Operations Research, 242*, 195–206.

Guan, J., Manikas, J., & Boyd, L. H. (2019). The International Journal of Production Research at 55: A content-driven review and analysis. *International Journal of Production Research, 57*, 4654–4666.

Hossain, M., Patras, A., Barry-Ryan, C., Martin-Diana, A., & Brunton, N. (2011). Application of principal component and hierarchical cluster analysis to classify different spices based on in vitro antioxidant activity and individual polyphenolic antioxidant compounds. *Journal of Functional Food, 3*(3), 179–189.

Johnson, D. S. (1993). Random starts for local optimization. In *DIMACS workshop on randomized algorithms for combinatorial optimization*.

Kao, C. (2008). The authorship and country spread of operation research journals. *Scientometrics, 78*, 397–397.

Katz, J. S., & Martin, B. R. (1997). What is research collaboration. *Research Policy, 26*, 1–18.

Kaufman, L., & Rousseeuw, P. J. (1990). *Finding groups in data: An introduction to cluster analysis*. Wiley.

Kim, M. C., Zhu, Y., & Chen, C. (2000). How are they different? A quantitative domain comparison of information visualization and data visualization. *Scientometrics, 107*(1), 123–123.

Kleinberg, J. (2002). Bursty and hierarchical structure in streams. In: *KDD '02: Proceedings of the eighth ACM SIGKDD international conference on Knowledge discovery and data mining* ((pp. 91–101). ACM.

Kumar, R., Novak, J., & Tomkins, A. (2006). Structure and evolution of online social networks. In: *KDD '06*.

Laengle, S., Merigójm, J. M., Modak, N. M., & Yang, J. B. (2020). Bibliometrics in operations research and management science: a university analysis. *Annals of Operations Research, 294*, 769–813.

Liao, H., Tang, M., Li, Z., & Lev, B. (2018). Bibliometric analysis for highly cited papers in operations research and management science from 2008 to 2017 based on Essential Science Indicators. *Omega, 88*, 223–236.

Liu, D., & Stewart, T. J. (2004). Object-oriented decision support system modeling for multicriteria decision making in natural resource management. *Computers and Operations Research, 31*(7), 985–999.

Luxburg, V. U. (2007). A tutorial on spectral clustering. *Statistics and Computing, 17*, 395–416.

Manikas, A., Boyd, L., Pang, Q., & Guan, J. J. (2019). An analysis of research methods in IJPR since inception. *International Journal of Production Research, 57*, 4667–4675.

Merigó, J., & Yangj, B. (2017). A bibliometric analysis of operations research and management science. *Omega, 73*, 37–48.

Mirjalili, S., Gandomi, A. H., Mirjalili, S. Z., Saremi, S., Faris, H., & Mirjalili, S. M. (2017). Salp Swarm Algorithm: A bio-inspired optimizer for engineering design problems. *Advances in Engineering Software, 114*, 163–191.

Newman, M. E. J. (2003). The structure and function of complex networks. *SIAM Review, 45*(2), 167–256.

Newman, M. E. J. (2006). Modularity and community structure in networks. *Proceedings of the National Academy of Sciences of the United States of America, 103*, 8577–8582.

Porter, M. E. (1990). *The competitive advantage of nations*. The Free Press.

Rau, E. P. (2005). Combat science: The emergence of operational research in World War II. *Endeavour, 29*, 156–161.

Romero-Silva, R., & Marsillac, E. (2019). Trends and topics in IJPR from 1961 to 2017: A statistical history. *International Journal of Production Research, 57*, 4692–4718.

Saaty, T. L. (1988). What is the analytic hierarchy process? In G. Mitra, H. J. Greenberg, F. A. Lootsma, M. J. Rijkaert, & H. J. Zimmermann (Eds.), *Mathematical models for decision support* (pp. 109–121). Springer.

Shang, G., Saladin, B., Fry, T., & Donohue, J. (2015). Twenty-six years of operations management research (1985-2010): authorship patterns and research constituents in eleven top rated journals. *International Journal of Production Research, 53*, 6161–6197.

Shukla, P. K., & Deb, K. (2007). On finding multiple Pareto-optimal solutions using classical and evolutionary generating methods. *European Journal of Operational Research, 181*(3), 1630–1652.

Small, H. (1973). Co-citation in the scientific literature: A new measure of the relationship between two documents. *Journal of the American Society for Information Science, 24*(4), 265–269.

Van Waltman, L., Eck, N., & Noyons, E. (2010). A unified approach to mapping and clustering of bibliometric networks. *Journal of Informetrics, 4*(4), 629–635.

White, H., & Mccain, K. W. (1989). Bibliometrics. *Annual Review of Information Science and Technology, 24*, 119–186.

Zupic, I., & Cater, T. (2015). Bibliometric methods in management and organization. *Organizational Research Methods, 18*(3), 429–472.

# Economic Resilience and Foreign Development Aid: Lessons from Sub-Saharan Africa

**Somar Al-Mohammad, Ammar Jreisat, Mourad Messaadia, and Audil Rashid Khaki**

**Abstract** The main goal of this chapter is to provide an overview of the role of foreign development aid in advancing the economic and social welfare of sub-Saharan African countries. The development aid and donations are described as the main economic stabilizers and growth engines in sub-Saharan Africa over the last 50 years. The empirical analysis in this chapter employs the predictive machine learning (ML) models and utilizes the powerful ML algorithms family to measure the impact of foreign development aid on main socioeconomic indicators of 48 African countries between 2000 and 2020. The results in this chapter reveal that African countries are still dependent on denotations and foreign aid in promoting economic growth and in subsidizing health and education sectors. These outcomes suggest that the African countries need to adapt comprehensive strategies to reform their economies and institutions in order to reap the utmost benefits from foreign aid and donations. The outcomes of this study will also help in understanding and prognosticating the resilience of African economies to potential decrease in foreign development aid flows in future.

S. Al-Mohammad · A. R. Khaki
College of Business Administration, American University of the Middle East, Kuwait, Kuwait
e-mail: Somar.Al-Mohamad@aum.edu.kw; Audil.Rashid@aum.edu.kw

A. Jreisat (✉)
Department of Economics and Finance, College of Business Administration, Sakhir, Kingdom of Bahrain
e-mail: abarham@uob.edu.bh

M. Messaadia
Department of Marketing and Management, College of Business Administration, Sakhir, Kingdom of Bahrain
e-mail: mmessaadia@uob.edu.bh

© The Author(s), under exclusive license to Springer Nature Switzerland AG 2022
H. Masri (ed.), *Africa Case Studies in Operations Research*, Contributions to Management Science, https://doi.org/10.1007/978-3-031-17008-9_9

# 1   Introduction

Over the last few decades, policy makers, academics, and research scholars alike have heated the discussion over the impact of foreign development aid on the economic and social (social-economic) indicators as well as the role of aid on poverty alleviation in sub-Saharan African countries. The foreign development aid can be defined as the international transfer of funds from governments and institutions of multilateral assistance, such as the International Monetary Fund and The World Bank, to countries that are deemed to be standing in need for assistance to conform the slumping economic situations and reduce poverty levels (Moyo, 2009). In retrospect, development aid became a key element of the international foreign policy and economic system since the evolution of Breton Wood agreement in 1914 that aimed at establishing liberal capital markets and strengthening economic ties among developed and developing countries (Todaro, 1977). According to the Organization for Economic Co-operation and Development (OECD), foreign aid represents financial and development donations and assistance granted to countries that are listed as least development. The African countries have started receiving comprehensive development aid officially in the 1960s when the African continent witnessed a huge wave of national liberalization from colonial powers (Sanjeev, 2006). Since then, the African continent has been marked by its insatiable desire to and reliance on financial aid where it received more than one trillion US dollars in the last 50 years in forms of assistance, donations, and development aid (Moyo, 2009). Despite the commitment made by majority of African government in the United Nations World Summit of 2005 to adopt national development strategies and reform programs, these countries are still classified as under-developed and they depend on donations and financial assistance to provide basic services such as health, education, electricity, and roads to their populations. In fact, the development aid has failed to achieve the poverty alleviation goals as 75% of the world's poor population is located in Africa, and this in turn questions the effectiveness of role played by foreign development aid in advancing standards of living and spurring economic growth in the entire continent. Figure 1 demonstrates the geographic location of African continent.

In fact, a large group of policy makers and economists believed initially that development aid is synonymous to economic prosperity and social welfare as they supplement national resources and ameliorate economic growth and economic resilience of recipient countries. Regrettably, little or no contribution of development aid has been witnessed in terms of political and socioeconomic development (Ibrahim, 2017). Furthermore, it has been claimed by many research scholars that the African continent has become more aid-dependent due to the lack of efficient employment of foreign flows. The utilization of foreign assistance in economic activities and development project became stifled by the reckless and irresponsible strategies adopted by political elite in majority of recipient countries. Moreover, few economists contend that foreign aid caused more harm and distortion than

**Fig. 1** The African continent on world map. Source: Map prepared by authors

socio-economic development in recipient countries in Africa despite the good intentions of the donors.

Figure 2 illustrates the amount, in US dollar, of foreign development aid received by majority of African countries in 2019. It can be clearly noticed that African countries have received a considerable amount of aid in 2019, where it reached up to $4.5 billion for Ethiopia which represents the largest recipient of foreign development aid.

The main goal of this chapter is to provide an intimate knowledge of the role of foreign aid in advancing the economic and social welfare indicators in sub-Saharan African countries. The contribution of this chapter to the existing literature is threefold: First, the analysis in this chapter provides a comprehensive summary of the role of foreign development aid flows to Africa in advancing socio-economic indicator in the last 20 years, and for a sample of 48 countries of the African continent. Second, the model utilized in this chapter is based on the machine learning (ML) algorithms and it is claimed to be the best predictive modeling approach and it enables for highlighting and employing powerful modulation of variables through reliable and proper regression analysis. Third, the outcomes of this study will enable the policy makers to assess and predict the expected resilience of African economies to shocks in financial aid flows due to future and ongoing crises (such as the current Ukrainian crisis). The remainder of this chapter is organized as follows: part two presents the review of literature, part three demonstrates the data description section, part four demonstrates the method of the study, part five includes the results and analysis, and part six encompasses conclusion and future recommendations.

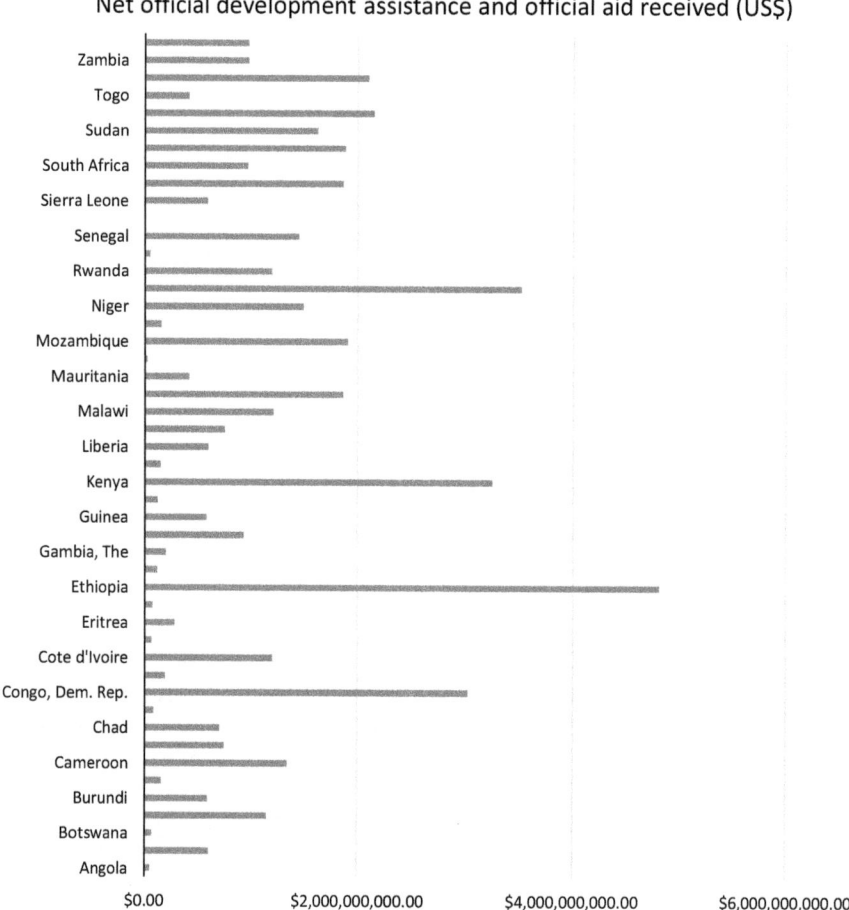

**Fig. 2** Net official development assistance received by African countries in 2019. Source: Prepared by the Authors using the World Bank Database

## 2 Literature Review

Over the last decade, a significant strand of literature has emerged to investigate the role of foreign development aids in boosting economic growth and alleviating poverty in the African continent. These attempts were propelled mainly by the fact that despite the huge amount of development aids received by African countries in the last 50 years, however these countries still account for half of Least Developed Countries (LDC) as per the UNDP Human Development Report in 2014.

The previous empirical works on the impact of foreign development aid on African countries did not reach a concord on whether aid and donation flow have spurred the main economic and financial growth engines, or they had negligible effects. In this regard, a group of studies measured the impact of foreign aid flow on social and macroeconomic indicators in Africa and came out with positive casual relationships. For instance, an early study by Carloos (1998) contended that by the end of the previous millennium, Botswana was able to shift from the heavy reliance on foreign development aids which accounted for more than half of government budget to least dependent on external aids flow and donations, and this was mainly due to the rational and efficient utilization of received aid in core infrastructure projects such as roads, education, training, and technology adoption. These developments have led to higher proportional economic growth rates ensued also as a result of FDI flows later on. In the same line, Burnside and Dollar (2000), Hansen and Tarp (2000), Morrissey et al. (2014), Dalgaard et al. (2004), and Gomanee et al. (2005) found positive relationship among foreign development aid flow and macroeconomic development and stability, they argued that the key factors for efficient distribution and utilization of foreign aid were improving modern technology adaptation, enhancing domestic human resources and saving, and stabilizing the currency exchange rate. Juselius et al. (2014) measured the long-run impact of aid flows on macroeconomic and investment indicators of 36 African countries, they found that development aid exhibited significant positive effect on either, investment, GDP or on both indicators in 27 countries. More recently, Bigsten (2018) asserted that Zambia has significantly benefited from the foreign aid assistance in the last 10 years (mainly from major donors such as China, Japan, and the USA). That despite the thorny issues pertinent to institutional inefficiency and political instability, the country was able to direct the aids flows into gain impetus in macroeconomic and foreign direct investments flows. The outcomes of the literature presented in this section are in line with the across-continent study performed by Ekanayake and Chatrna (2010) who evaluated the efficiency of 83 recipient nations, in Africa, Asia, Caribbean, and Latin America, in utilizing the development donations and aids. The outcomes suggested that African countries have experienced a positive impact of development aids on their GDPs, while the recipient countries in Asia, Latin America, and the Caribbean have been subject to adverse effects on economic growth rates.

A handful number of studies, on the other hand, had assiduously explored the efficiency utilization of foreign aids in Africa and found scant or no clues on positive impacts of the development aids on economic and financial growth as well as poverty alleviation. On the contrary of previous literature on positive impacts of development aids on African economies, it has been widely observed by a number of studies that aid assistance has incapacitated economic growth and spurred corruption across the majority of recipient nations in Africa. In retrospect, Boone (1996) and Easterly (2018) found that due to the absence of minimum levels of governance, institutions quality, political rights, and economic freedom in many African countries, the development aid could foster corruption, increase poverty, and prohibits the recipient countries from taking opportunities arise in global economy. The proponents of this view base their arguments on the outcomes of several empirical works that concluded that the foreign development aid had negative consequences on most recipient countries in Africa. For instance, Ndikumana and Boyce (2011) contended that despite the $3.6 billion received by Angola in forms of aids and donations during the 1990s, the country had tremendously suffered from lack of governance and transparency regarding the national budget expenditure and it remained to be one of the poorest countries in the world. Berry (2016) argued that the foreign development aids received by Tanzania originated from more than 100 donors including developed countries and international development institutions. Based on the results of the study, however, it appears that aid has not been realized to its full potential because of a combination of the incorrect direction and orientation of aid, the lack of human resource capacity, and the inability to mobilize the effort for efficient resource allocation. The same outcomes were revealed by United Nations Development Program (UNDP) in 2014, where Ethiopia was ranked as 173 out of 183 of least development countries despite being one of the main recipients of international donations. This was mainly due to lack of democratic foundation and the outbreak of profound corruption. Similarly, Appiah et al. (2016) claimed that despite the noticeable economic development observed in Ghana during the 1980s, the absence of full-fledged implementation of reform programs and tenuous efficiency of economic and fiscal policies have negatively affected the proper employment of development aid received in the 1990s. These deficiencies revealed serious economic problems such as unstable exchange rate, inflation volatility, high level of national indebtedness, and sluggish economic situation. Ibrahim (2017) found that development aids created socio-economic difficulties in Somalia such as the high violent rates, corruption, and weakened the Somalia's economy. Moreover, 70% of foreign aids allocated for economic development between 2009 and 2010 were embezzled. Finally, Rouis (2019) reported that in 2018, up to 16% of Senegal's GDP was accounted for by foreign development donations. Nevertheless, the economic outlook of Senegal remained shrouded in pessimism due to erroneous allocation of funds and lack of systematic planning and leadership. It can be clearly noticed that the relationship between foreign development aids and the main macroeconomic, financial, and social indicators depends to large extent on important factors such as the quality of the beneficiaries' institutions, national economic policies, social stability, political conditions, and the existence of competent laws that guarantees the appropriate and efficient allocations of foreign aids (McGillivary et al., 2006).

Based on previous empirical studies, there has been a notable lack of empirical studies assessing the effects of foreign aid on economic growth on the African continent. These studies estimate the efficiency of major African recipients in utilizing foreign contributions for economic growth. There has been an adverse impact on a wide range of socio-economic indicators simultaneously and multidimensionally as a result of these aid flows. However, the findings are not unified among researchers. Hence, this chapter intends to utilize the powerful machine learning (ML) algorithms family to assess the impact of foreign development aid on the main socioeconomic indicators of 48 African countries between 2000 and 2020.

## 3 Data Description

As the primary objective of this study is to evaluate the impact of foreign development aid on the economic development in African countries, the focus thus is on the impact of development aid particularly on education, health, food, and economic growth in 48 African countries during the period from 2000 to 2020. The current expenditures on education and health, separately, as a percentage of GDP are chosen as a proxy for the quality of education and health services, respectively. Food imports as a percentage of total merchandise imports represents the level of food security in the continent. The GDP growth percentage is representing the economic growth. Moreover, the inflation and employment indicators are selected as control variables to reflect the socio-economic development in the African countries. Lastly, the Net official development assistance and official aid received accompanied by the Net bilateral aid flows from DAC donors indicate the amounts of foreign development aids flowed to the continent between 2000 and 2020. For detailed information about the variables utilized in this study, Table 1 below illustrates the name, description, source, and indication of the variables employed in this study.

## 4 Method of Study

In this study, we apply the machine learning (ML) algorithm framework in order to produce the most efficient prediction model to quantify the impact of foreign aid on major socio-economic variables in 48 African countries between 2000 and 2020. The ML approach includes several approaches of supervised learning such as Support Regression Vector (SVR) and Linear Regression (LR) to embed the dataset into best predictive forecast model. The generation of the best predictive ML model is accomplished through different steps including data cleaning and organizing, training dataset identification, running training dataset through identified iterative algorithm to generate best ML predictive model, and finally employing the best predictive methodology on new dataset to improve the efficacy of the dynamic iteration process.

The ML process describes the observations by a set of predictor variables such as $a_1, a_2, \ldots, a$, each predictive variable $a_{i,\,j}$ denotes for individual $i$ by $a_j$. The ML process initiates a matrix A for vectors of predictive variables with $m \times n$ dimensions. This matrix is represented as follows:

$$A = \begin{bmatrix} a_{11} & \cdots & xa_{1n} \\ \vdots & \ddots & \vdots \\ a_{m1} & \cdots & a_{mn} \end{bmatrix} \tag{1}$$

**Table 1** List of variables utilized in this study

| Variable | Topic | Source | Description | Mnemonic |
|---|---|---|---|---|
| Current health expenditure (% of GDP) | Health: Health systems | World Health Organization | Level of current health expenditure expressed as a percentage of GDP includes healthcare goods and services consumed during each year | HEALTH |
| Employment to population ratio, 15+, total (%) (modeled ILO estimate) | Social Protection & Labor: Economic activity | International Labor Organization, | Employment to population ratio is the proportion of a country's population that is employed | EMP |
| Food imports (% of merchandise imports) | Private Sector & Trade: Imports | United Nations Statistics Division. | Food comprises the commodities of food, beverages, oils, and fats | FOOD |
| GDP growth (annual %) | Economic Policy & Debt: National accounts: Growth rates | World Bank | Annual percentage growth rate of GDP at market prices based on constant local currency | GROWTH |
| Government expenditure on education, total (% of GDP) | Education: Inputs | UNESCO | General government expenditure on education as a percentage of GDP | EDU |
| Inflation, consumer prices (annual %) | Financial Sector: Exchange rates & prices | International Monetary Fund | Inflation as measured by the consumer price index reflects the annual percentage change in the cost to the average consumer of acquiring a basket of goods and services | INF |
| Net bilateral aid flows from DAC donors, Total (current US$) | Economic Policy & Debt: Official development assistance | Development Assistance Committee of the Organization for Economic Co-operation and Development. | Net bilateral aid flows from DAC donors are the net disbursements of official development assistance (ODA) or official aid from the members of the Development Assistance Committee (DAC) | DACAID |
| Net official development assistance and official aid | Economic Policy & Debt: Official | Development Assistance Committee of the Organization for Economic | Net official development assistance (ODA) consists of disbursements of | NODA |

(continued)

**Table 1** (continued)

| Variable | Topic | Source | Description | Mnemonic |
|---|---|---|---|---|
| received (current US$) | development assistance | Co-operation and Development. | loans made on concessional terms (net of repayments of principal) and grants by official agencies of the members of the Development Assistance Committee (DAC) | |

In matrix A, $m$ indicates the number of records while $n$ the number of features for each observation, the observation in the training dataset is defined by the above-mentioned predictor variables of $a_1, a_2, \ldots, a$. One of the main powerful features of the ML method is that it accounts for and detects the existence of correlation among studied variables, consequently it removes the highly correlated variables in order to generate unbiased results. In this study, the independent variables of Net bilateral aid flows from DAC donors (DACAID) and the Net official development assistance and official aid received (NODA) are represented by $a_{j1}$ and $a_{j2}$, respectively, are used to generate a prediction of dependent or output variable $b$ using the observations in the training dataset. The independent or output variables in this study are represented by a vector of HEALTH, FOOD, EDU, and GROWTH variables, in addition to the selected controlling variables of employment (EMP) and inflation (INF) rates. The set of output variables will be represented by a vector of size $m$.

$$B = \begin{bmatrix} b_1 \\ b_2 \\ \ldots \\ b \end{bmatrix} \tag{2}$$

The goal of machine learning is to determine a function between a quantitative variable B (output or dependent variable) and $n$ different dependent predictor variables or inputs $a_1, a_2, \ldots, a_n$. The relationships among the set of dependent variables denoted by $B$ and independent variables denoted by $A$ are measured by the following function:

$$b = f(A) + \varepsilon \tag{3}$$

The $f$ denotes a fixed and unknown function that ties $(a_1, a, \ldots, a_n)$ to B, $\varepsilon$ is the error term defined as a random distribution with a mean ($\mu = 0$) and independent of A. The model was verified with the training data for the efficiency and performance of the model. The final model generated by the ML approach described above is described below:

$$
\begin{bmatrix}
\text{HEALTH} \\
\text{FOOD} \\
\text{GROWTH} \\
\text{EDU}
\end{bmatrix}
=
\begin{vmatrix}
C + \alpha_1 \text{NODA} + \alpha_2 \text{EMP} + \alpha_3 \text{INF} + \varepsilon \\
C + \alpha_1 \text{DACAID} + \alpha_2 \text{EMP} + \alpha_3 \text{INF} + \varepsilon
\end{vmatrix}
\tag{4}
$$

The model in Eq. (4) represents the best predictive multiple regression suggested by the ML. The left side of the equation includes a matrix of dependent variables for health, food, economic growth, and education indicators in the group of 48 African countries. *NODA* and *DACAID* in eq. 4 demonstrate the foreign development aid flows to these countries from 2000 to 2020. Also, the employment and inflation rates are included as controlling variables, and $\varepsilon$ represents the error terms.

## 5    Results and Discussion

The impacts of the foreign development aid (represented by Net bilateral aid flows from DAC donors and Net official development assistance and official aid received) on the socio-economic development factors (represented by health, food, economic growth, and education indicators) on African countries are based on the outcomes of the best predictive multiple regression suggested by the ML algorithms. Table 2 demonstrates the contribution of Net bilateral aid flows from DAC donors (DACAID) on socio-economic indicators. Results in Table 2 illustrate that the $p$ values of the multiple regressions indicate a positive impact of DACAID on Education, Economic Growth, and Health indicators with 5% significance level, however the impact of the Food is found to be negative and significant at a level of 10%. These results imply that the increase in official development aid flow from

**Table 2**  Regression analysis: DACAID and socio-economic development

|  | Dependent variable | | | |
|---|---|---|---|---|
|  | Education | Economic growth | Food | Health |
| C | 5.391* | 2.935** | 22.936*** | 5.401*** |
|  | (1.104) | (1.246) | (5.068) | (1.297) |
| DACAID | 8.92e-11** | 1.66e-09** | −6.89e-09* | 1.10e-09** |
|  | (3.46e-11) | (6.42e-10) | (3.96e-09) | (4.02e-10) |
| EMP | −0.021 | 0.016 | 0.023 | 0.002 |
|  | (0.019) | (0.021) | (0.085) | (0.022) |
| INF | −0.016 | −0.073*** | −0.006 | 0.02 |
|  | (0.016) | (0.018) | (0.072) | (0.018) |
| R-squared | 0.068 | 0.309 | 0.232 | 0.039 |
| AIC | 192.5 | 204.1 | 338.8 | 207.9 |
| BIC | 199.9 | 211.6 | 346.3 | 215.4 |

Numbers in the parenthesis indicate standard errors
***, **, and * indicate significance at 99%, 95%, and 90% confidence intervals, respectively

**Table 3** Regression analysis: NODA and socio-economic development

| | Dependent variable | | | |
|---|---|---|---|---|
| | Education | Economic growth | Food | Health |
| C | 5.4052*** | 3.2417** | 21.811*** | 5.274*** |
| | (1.1111) | (1.215) | (5.141) | (1.111) |
| NODA | 2.98e-10** | 1.34e-09*** | −5.71e-9*** | 3.99e-10** |
| | (1.16e-10) | (4.22e-10) | (1.79e-09) | (1.65e-10) |
| EMP | −0.022 | 0.008 | 0.042 | 0.005 |
| | (0.019) | (0.021) | (0.088) | (0.022) |
| INF | −0.016 | −0.071*** | −0.03 | 0.019 |
| | (0.015) | (0.017) | (0.07) | (0.018) |
| R-squared | 0.068 | 0.353 | 0.22 | 0.047 |
| AIC | 192.5 | 201.0 | 339.5 | 207.5 |
| BIC | 199.9 | 208.5 | 347.0 | 215.0 |

Numbers in the parenthesis indicate standard errors
***, **, and * indicate significance at 99%, 95%, and 90% confidence intervals, respectively

Development Assistance Committee (DAC) to African countries leads to more expenditure on Education, Health sectors, as well as enhancing the GDP of the recipient counters. On the contrary, the increase in the aid leads to slight decrease in Food imports as a percentage of total merchandise imports. However, the coefficients of the regressions in the four panels in Table 2 indicate a low or negligible impact of DACAID on Education, Economic Growth, Health, and Food indicators. The same outcomes can be noticed from Table 3 that displays the effects of Net official development assistance and official aid (NODA) on the socio-economic development indicators. The second development indicator (NODA) seems to have positive impact on Education, Economic Growth, and Health indicators, whereas the impact on the level of food exports is negative with 1% significance level.

The outcomes in this study are in line with the previous findings of Wangwe (2006), Juselius et al. (2014), Alemu and Lee (2015), EIU (2016), and Fuhrer (2017) who provided a clue for the positive impact of foreign development aid on the economic growth in many African countries. In the same line, the results in this study affirm the findings of the World Bank report (2013) on the impact of development aid on the education infrastructure in Africa, for instance the foreign development aid has enabled Angola to build up more than 110 schools starting from 2011 (Alden, 2014). Regarding the positive impact of donations on the health sector, the outcomes in this study are somewhat plausible since many African countries were able to increase their expenditure on hospitals and health services using development aid. For instance, in the last 10 years aids comprised up to 35% of total expenditure on the health sector in Ghana (Durauraj et al., 2016), also Easterly (2018) claimed that Angola utilized $180 million of foreign donations in hospitals renovation and improvements during 2017–2018. The results pertinent to the negative impact of development aid indicators on food imports might be explained by the lack of technological adoption in the agricultural sector in the majority of African countries, however these findings do not seem to be in parallel with the outcomes of the World Bank report (2015) which affirmed on the importance of aid to most African countries in rehabilitating large-scale farming lands to boost the agriculture and food productivity.

The results in this study, in general, suggest that the increase in foreign aid flow to the African continent will lead to slight enhancement in socio-economic indicators, with an exception of food imports. These outcomes are consistent with the unanimity of main strand of previous literature, where the African countries are found to remain dependent on denotations and foreign aids in providing the main services for their societies. However, the most noticeable outcome, could be, is the very low scale and impact the development aid exerts on these variables. This could be explained by the lack of full-fledged management of aid accompanies with the long history of social unrest and political instability. Also, the African countries do not seem to be an exception among the third world countries when it comes to high levels of political and institutional ineffectiveness and corruption. These factors combined lead to waste of huge amounts of foreign aid flows, hence the African countries should improve their efficiency, utility, and management of foreign development aid flows to enhance the food safety and security in the region, especially in the wake of the ongoing Ukrainian crisis that led to sharp increase in prices of wheat.

# 6 Conclusion and Recommendations

The aim of this study is to assess the impact of foreign development aid on main socio-economic indicators in African countries. The study employs the multiple regression method suggested by the machine learning (ML) framework in order to produce the most efficient prediction model. The results reveal that foreign aid positively affects the health, education, and economic growth variables, however the impact on food imports was negative and negligible. This could be explained by the lack of full-fledged management of aid accompanies with the long-history of social unrest and political instability. In view of the results of this study, some lessons and future directions for policy and regulation frameworks can be developed and revisited. The results in this study reveal that the African countries are in need for a comprehensive reform to reap the best outcomes from foreign aid flowing to the region. An important starting point of the comprehensive reform could be changing the African governments perception and view of development aids as a source of national income, but instead they should efficiently allocate foreign donation to advance the technological infrastructures in their countries. Also, the technological adaption by services and agricultural sectors should be in the top of the policy makers agendas. Moreover, the international aids flow to African countries needs to be parallel to an enhanced international collaboration with countries that succeeded pushing their technological frontiers. On top of the above, the entrepreneurship initiatives and private sector developments, as well as the legal infrastructure and tax laws need to be promoted. Finally, the African Continental Free Trade Area agreed in 2018 needs to be boosted and activated to enhance the investment and trade income in these countries.

# References

Alden, C. (2014). *Mapping official development assistant in Africa: A synthesis analysis of Angola, Zambia and Zimbabwe*. Institute of Development Dialogue.

Alemu, M. A., & Lee, J.-S. (2015). Foreign aid on economic growth in Africa: A comparison of low and middle-income countries. *SAJEMS NS, 184*, 449–462.

Appiah, K., Pual, J., Forster, S., Eric, A., & Twerefou, D. (2016). The effect of foreign aid on economic growth in Ghana. *African Journal of Economic Review, 4*(2), 114–126.

Berry, L. (2016). *The foreign aid sector in Tanzania*. Scandinavian Institute of African Studies.

Bigsten, A. (2018). *An evaluation of foreign aid in Zambia*. SASDA.

Boone, P. (1996). Politics and the effectiveness of foreign aid. *European Economic Review, 40*(2), 289–329.

Burnside, C., & Dollar, D. (2000). Aid, politics and growth. *The American Economic Review, 90*(4), 847–868.

Carloos, J. (1998). *Aid effectiveness in Botswana*. Economic Research Bureau.

Dalgaard, C., Hansen, H., & Tarp, F. (2004). On the empirics of foreign aid and growth. *African Economic Journal, 114*(496), 191–216.

Durauraj, V., Selassi, D., & Kirigia, J. (2016). *Ghana's approach to social health protection*. Background Paper for the 2017 World Health Report. World Health Organization.

Easterly, W. (2018). *Chinese investments in Angola: An economic perspective*. African Labor Research Network.

EIU. (2016). *Country report: Angola*. Economist Intelligence Unit (EIU).

Ekanayake, E., & Chatrna, D. (2010). The effect of foreign aid on economic growth in developing countries. *Journal of International Business and Cultural Studies, 3*, 140–155.

Fuhrcr, II. (2017). *The story of official development assistance: A story of official development assistance in Angola* (pp. 121–156). OECD.

Gomanee, K., Girma, S., & Morrissey, O. (2005). Aid and growth in sub-Saharan Africa: Accounting for transmission mechanisms. *Journal of International Development, 17*(8), 1055–1075.

Hansen, H., & Tarp, F. (2000). Aid effectiveness disputed. *Journal of International Development, 12*, 375–398.

Ibrahim, B. A. (2017). *How foreign aid hurts famine relief in Somalia*. Foundation for Economic Education.

Juselius, K., Møller, F. N., & Tarp, F. (2014). The long-run impact of foreign aid in 36 African countries: Insights from multivariate time series analysis. *Oxford Bulletin of Economics and Statistics, 76*(2), 50–76.

McGillivary, M., Feeny, S., Hermes, N., & Lensink, R. (2006). Controversies over the impact of development aid: It works: It doesn't; it can, but that depends. *Journal of International Development, 18*(7), 1031–1050.

Morrissey, O., Carisma, B., & Lowder, S. (2014). *Financial resource flows to Angola's agriculture sector: A review of data on foreign direct investment, official development assistance, and government spending*. FAO, ESA Working Paper No. 11–19.

Moyo, D. (2009). Why foreign aid is hurting Africa. *The Wall Street Journal, 21*, 1–5.

Ndikumana, L., & Boyce, J. (2011). *Africa's odious debts*. Zed Books.

Rouis, T. (2019). *Stabilization, partial adjustment and stagnation of Senegal. Adjustment in Africa: Lessons from country case studies*. World Bank.

Sanjeev, G. (2006). *Macroeconomic challenges of scaling up aids to Africa: A checklist for practitioners*. International Monetary Fund.

Todaro, M. P. (1977). Economics for a developing world. *Journal of International Development,* *15*(3), 425–438.

Wangwe, S. (2006). Foreign aid in Africa: Role, experiences and challenges. In *Background paper* *for the AfDB conference in Tunis*, 22 November 2006.

World Bank. (2013). *Lifelong learning in global knowledge economy: Challenges for developing* *countries*. A World Bank Report.

World Bank. (2015). *World development report 2015: Agriculture for development*. The World Bank.

# Fintech Adoption for Poverty Alleviation in African Countries: Application of Supervised Machine Learning Approach

**Audil Rashid Khaki, Mourad Messaadia, Ammar Jreisat** (iD),
**and Somar Al-Mohammad**

**Abstract** Financial Inclusion is believed to play an important role in spurring economic growth, enterprise development, reducing income inequalities & child labor, enabling women empowerment, and poverty alleviation. Over the last decade, the adoption of technology and innovative approaches to the dispensation of financial products and services have witnessed leapfrogging beyond the conventional frontiers of financial inclusion. African economies are characterized by weakly developed financial systems, poor infrastructure, inadequate brick-and-mortar banking outreach, and extreme poverty. The current study attempts to examine and revisit the relationship between financial inclusion, digital/technological penetration, and poverty alleviation in the African countries as digital financial inclusion platforms evolve. The study employs a supervised machine learning (ML) approach to identify the best predictive multiple regression model and the best combination of technological, financial inclusion, and fintech variables to evaluate the impact on poverty alleviation in the region. The results suggest that among the technological variables, mobile phone outreach has a considerable impact on poverty alleviation. Among the conventional financial inclusion variables, ATM coverage and access to formal credit have been found to significantly contribute to the reduction of poverty in the region while the brick-and-mortar banking outreach has no impact on reducing poverty. Among fintech variables, the breadth of mobile money agents, mobile money transactions, and the use of mobile and internet to access accounts substantially contribute to poverty alleviation while the other fintech variables have no

A. R. Khaki · S. Al-Mohammad
College of Business Administration, American University of the Middle East, Kuwait, Kuwait
e-mail: Audil.Rashid@aum.edu.kw; Somar.Al-Mohamad@aum.edu.kw

M. Messaadia
Department of Marketing and Management, College of Business Administration, Sakhir, Kingdom of Bahrain
e-mail: mmessaadia@uob.edu.bh

A. Jreisat (✉)
Department of Economics and Finance, College of Business Administration, Sakhir, Kingdom of Bahrain
e-mail: abarham@uob.edu.bh

© The Author(s), under exclusive license to Springer Nature Switzerland AG 2022          197
H. Masri (ed.), *Africa Case Studies in Operations Research*, Contributions to Management Science, https://doi.org/10.1007/978-3-031-17008-9_10

significant impact. The results bear important policy implication as it is imperative to consider the local landscape while designing the comprehensive financial inclusion policy as well as in embracing the digital financial inclusion framework, given the regulatory challenges associated with fintech and digital approaches.

# 1 Introduction

Financial Inclusion has been defined as unrestrained universal access to formal sources of financial products and services for all sections of society and business entities. It essentially entails access to affordable financial products and services, including savings, credit facilities, insurance, consulting advice, and other investment products. While financial inclusion broadly covers access to all, it mainly emphasizes the need to reach out to the traditionally unbanked or underbanked sections of the society. The conventional financial system has struggled to extend the scope of its products and services to the so-called unbankable and underserved sections of the society, mostly due to poor infrastructure limiting the outreach of banks and/or non-viability of small-ticket banking transactions (Sapovadia, 2018). While the conventional approaches to expand the outreach of formal financial services have largely been unsuccessful, the recent innovations in digital innovation and technologies, particularly by the financial institutions have made both the financial and non-financial entities leapfrog beyond the traditional frontiers of the financial system. The emergence and adoption of digital solutions in the provision of financial services have undisputedly disrupted the financial inclusion landscape and powered the recent growth in financial inclusion across different regions of the world (Hotchkiss & Chuen, 2018). Despite huge leapfrog developments in expanding the outreach of financial products and services through the conventional "brick-and-mortar" approaches as well as the disruptive fintech approaches, some regions around the world, Africa in particular still remains to be considered deprived access to formal financial access, particularly due to poor road and telecom infrastructure. As could be seen in Fig. 1 below, most African countries have very low access to account ownership at a formal financial institution or a mobile money account with a mobile money agent.

Given the huge regional disparities in the provision of financial services and the enormous success of the pioneering mobile money (M-Pesa) model in Africa, the disruptive financial inclusion approaches appear to offer a promise toward achieving sustainable financial inclusion in the African region. Fintech offers a huge prospect for financial and non-financial institutions to provide financial services to traditionally excluded clients at an affordable price leveraging on the conventional approaches and offering innovative delivery channels through tech-driven platforms like mobile banking, etc. The emergence and integration of innovative digital delivery solutions in delivering financial services to the underserved and excluded population has given rise to and extended the scope of financial inclusion to digital financial inclusion. The world leaders, Jim Yong Kim, the president of the World

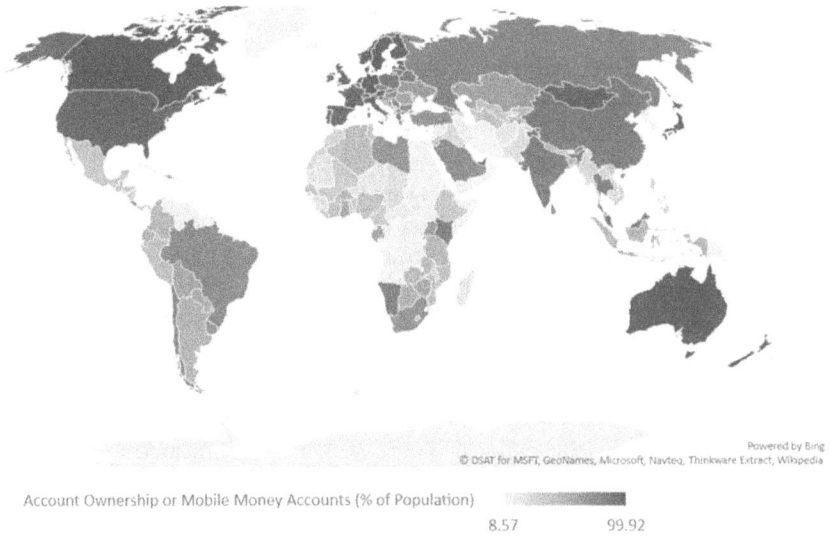

**Fig. 1** Account ownership and mobile money accounts (data from World Bank, map developed by authors)

Bank Group, in particular, are optimistic about the potential of digital financial inclusion in extending the universal access to financial access if the new technologies, innovative business models, and ambitious reforms are undertaken in a collaborative manner. Digital Financial Inclusion is defined as the deployment of cost-effective approaches and models with a wide range of products, services, and delivery models to reach the financially excluded and excluded sections of the society, delivered responsibly to the clients and sustainable for providers (GPFI, 2016; Lauer & Lyman, 2015). The key components of digital financial inclusion are the transaction platforms, devices, facilitating agents, and additional service providers. It is believed that the provision of digital financial inclusion could be transformational yet affordable to the segments of the population ignored by the traditional financial system, thereby, leading to economic empowerment, social capital formation, risk mitigation, and subsequently poverty alleviation.

Financial inclusion has seen a substantial shift in the last few years. From being a dichotomous division of included or not to being a multidimensional concept. This has allowed both service providers and policymakers to evaluate whether their products, practices, frameworks, and regulations are effective. Multidimensional financial inclusion includes the dimensions of access, usage, and quality and strives to improve not only the outreach of financial inclusion but also the quality and effectiveness of financial inclusion in impacting the socio-economic wellbeing of customers (AFI, 2013).

African region appears to be the most underserved region in terms of SME financing, besides individual financial access. The SME finance gap is the highest in the African region where almost 50% of the SMEs lack access to formal financing leading to a gap of more than $100 billion in credit for SMEs alone (Demirgüç-Kunt & Klapper, 2012). The acuteness of the funding gap further worsens as the size of the firm decreases resulting in higher failure rates and higher business risk for small businesses in the region. The traditional approaches to expanding the financial services have failed in Africa just like in other developing countries. It is, therefore, imminent for financial service providers to break beyond the conventional approaches to financial access by scaling up the transformational solutions built on the evolving technological platforms and reach out to the underserved population through mobile telephone platforms (mobile money) and other innovative approaches (Stein et al., 2013). The African region has assumed a pioneering position in the adoption of fintech solutions to financial inclusion and SME growth. Given the limited brick-and-mortar banking infrastructure in the region, the region has demonstrated great resilience in adopting fintech solutions in payments, savings, and credit processes, thereby, expanding the outreach of financial products and services beyond the conventional boundaries of the financial system. The technology-based fintech solutions have the potential to foster economic growth and alleviate poverty in the region if comprehensive and collective efforts are undertaken to build an efficient and sustainable financial infrastructure that breaks beyond the conventional financial inclusion frontiers. The current study is an attempt to understand the contribution of the latest developments in the multidimensional financial inclusion in the region on poverty alleviation by implementing the machine learning algorithm-based predictive model.

## 2 Literature Review

One of the key pillars of success in expanding the outreach of financial services is financial innovation. While there are competitive pressures in the finance industry that one would expect financial institutions to explore opportunities to offer cost-effective mechanisms and expand their markets, the evidence suggests that the financial institutions have not been so forthcoming in undertaking the innovation in products, services, and delivery approaches. Given the reluctance of conventional financial intermediaries to innovate, the financial innovation is mostly coming from non-financial quarters and new entrants (Allen et al., 2020; Beck, 2020). Digital financial inclusion has witnessed a huge growth in recent years with fintech providers extending into the provision of credit and mortgage lending too (Buchak et al., 2018; Fuster et al., 2019). Digital financial inclusion has not only brought about a transformation in the peer-to-peer transaction mechanisms but also allowed non-financial institutions and big-tech firms to participate and enhance the quality of financial products and services (Beck, 2020; Claessens et al., 2018). While initially limited to payment services, the digital financial inclusion platforms are expanded to push the access frontiers even further by offering credit and lending services (Beck et al., 2018). Moreover, the funding gaps in personal and SME lending left unaddressed by the conventional financial sector have been filled by peer-to-peer (P2P) lending platforms offered by tech firms and other new entrants.

The literature on the impact of digital financial inclusion on community welfare and enterprise growth has been rapidly expanding, though limited. Most of the research on digital financial inclusion is focused on Africa, though, due to the lack of brick-and-mortar infrastructure and its overwhelming adoption of fintech solutions. Claessens et al. (2018), for instance, argue that fintech credit has the potential to contribute more to the economies with the less competitive banking system, such as Africa, both in adoption and effectiveness of fintech mechanisms. Similarly, Faye and Triki (2013) believe that with a growing mobile subscriber base in Africa and the absence of the "brick-and-mortar" infrastructure, the potential of fintech and other technology-driven delivery platforms is enormous and could revolutionize the continent's financial inclusion and delivery ecosystem. While there is no general consensus on whether the bank-led approaches are more successful or the alternative finance approaches like mobile money, it is imperative that the inclusion approaches that challenge the existing mechanisms and disrupt the existing conventional frontiers are effective in the dispensation of financial services to the financially excluded segments of the population and small businesses alike.

Mbiti and Weil (2015) find that M-Pesa (mobile money) usage in Kenya is effective in weaning off customers from informal mechanisms and increasing the use of formal banking approaches. They further found that M-Pesa leads to an increase in the P2P transfers while also decreasing the cost of competing money transfer services. Kikulwe et al. (2013) report similar findings suggesting that the use of M-Pesa in Kenya has been instrumental in creating a net positive impact on the household welfare of small-scale farmers by reducing the risk constraints and promoting the commercialization of agriculture. The users were found to face lesser liquidity constraints, increase their farm input, expand marketing of their output, and subsequently have higher profits with higher household income. Wang and Fu (2021) argue that digital financial inclusion has a transformative capability in reducing the vulnerabilities of poor households in China and enabling them in mitigating shocks, improve agricultural productivity, stimulate enterprise growth, and promote non-farm employment. Similarly, Wieser et al. (2019) found that the introduction of mobile money in Uganda contributed to higher employment in the non-farm sector and an increase in the remittance transactions while on the contrary, Aggarwal et al. (2020) report that the adoption of mobile money reverts the employment from business to agriculture in Malawi, as well as, increases the rate of saving at a formal institution.

It is generally argued that access to finance results in the efficient employment of resources, particularly at the micro, small, and medium enterprise level increasing the productivity of agricultural, handicrafts, and other businesses, thereby, generating employment and reducing poverty (Churchill & Marisetty, 2020; Demirgüç-Kunt et al., 2015; Erlando et al., 2020; Inoue, 2018; Khaki, 2017; Khaki & Sangmi, 2016). Soyemi et al. (2020) report that financial inclusion has a significant positive impact on sustainable development, enterprise development, poverty alleviation, and household welfare. Mohammed et al. (2017) assert that financial inclusion has helped in reducing poverty considerably in sub-Saharan African countries; they found that the poor financially included derived a positive net wealth benefit from financial inclusion. Global Partnership for Financial Inclusion—GPFI (2016) asserts that poverty and inequality still are the challenges that need to be effectively addressed, particularly in the areas with inadequate road and telecom infrastructure. Digital Financial Inclusion is considered vital in achieving inclusive growth, reducing inequalities, alleviating poverty, and enterprise growth through capacity building and expanding the outreach of innovative financial delivery mechanisms. Suri and Jack (2016) report that the adoption of M-Pesa increased household consumption of Kenyan households and lifted 2% of the households out of poverty; they also found that the impact was particularly higher on women who moved out of their traditional agricultural occupation into business. Mushtaq and Bruneau (2019) report that when ICT and financial inclusion are packaged as digital financial inclusion and rolled out as sustainable financial products and services, it has the potential to accelerate economic growth, alleviate poverty, and reduce inequality. Tay et al. (2022) argue that the developing countries with a poor banking infrastructure have shown eagerness to embrace digital financial inclusion and have substantially contributed to poverty alleviation and economic development. Beck et al. (2018) found that small enterprises with higher productivity in Kenya have a high propensity of adopting M-Pesa mechanism and are positively affected in terms of business credit and enterprise growth.

While fintech is largely believed to expand the frontiers of financial inclusion, create synergies for inclusive development, poverty alleviation, economic growth, etc., it is also believed to create regulatory issues for artificial intelligence (AI) driven algorithms in fintech products and services (Philippon, 2019). Mehrotra (2019) argues that the emergence of fintech-based financial solutions despite incredibly enhancing the quality of products and services has created a parallel banking ecosystem and poses a serious challenge to the regulators. It is also believed that the technology-led financial inclusion approaches operating on artificial intelligence (AI) and machine learning (ML) frameworks may eventually optimize with a robo-prejudice and may not effectively reduce inequalities in the dispensation of financial services (Mehrotra, 2019; Philippon, 2019). The most prominent challenges around the emergence of fintech and its regulation are blurring boundaries of operation, the unharmonized scope of regulation, disintermediation, rapid scalability, and the lack of reliable information (World Bank, 2020a, 2020b).

## 3   Data Description and Methodology

The study employs a Machine Learning (ML) algorithm to generate the most efficient predictive model built on a certain training data algorithm. The approach is generally employed to improve the accuracy and efficiency of the predictive model depending upon the nature of a relationship and the type and volume of data employed. The approach employed for this study involves supervised learning with a well-defined data set and some understanding of how that data is classified with the goal of achieving the best predictive model. Supervised learning includes several approaches such as SVR (Support Regression Vector), ANN, Linear Regression, Decision tree, etc. The following approach has been adopted in generating the predictive ML model for our study.

- First, the data set is properly cleaned, organized, and labeled to be used as training set data.
- Second, the training data is identified, and based on the nature and volume of the data, select an algorithm to optimize the solution for a problem. For the current study, multiple regression was selected as a classifier algorithm.
- In the third step, the training data is run through the identified algorithm in an iterative process to identify the best predictive model. The algorithm, thus trained, is called the Machine Learning model.
- In the last step, the model is employed on new data to improve its efficiency in a dynamic iterative process.

The description of the variables employed in our adopted Machine Learning (ML) approach is provided in Table 1 below. The data for all the variables was downloaded from the World Bank Database for a period from 2011–2019 for 35 African Countries.

For Machine Learning models, any observation in the training set is described by a set of predictor variables $x_1, x_2, \ldots, x_n$. To make reliable predictions, a Machine Learning algorithm will need several observations to detect correlations between the data. These observations are called the Training Set. After the initial correlation analysis, some variables were found to be highly correlated for the model to generate reliable and unbiased results and were, therefore, dropped from further analysis. We denote the predictor variable $x_{i, j}$ of individual $i$ by $x_j$. The data for all the predictor variables are then stored in a Matrix denoted by $X$ with dimensions $m \times n$, with m the number of records and n the number of features for a given observation. For a machine learning algorithm to use all this data, it is represented in matrix form as below.

$$X = \begin{bmatrix} x_{11} & \cdots & x_{1n} \\ \vdots & \ddots & \vdots \\ x_{m1} & \cdots & x_{mn} \end{bmatrix}$$

In our work, for example $x_{j1}$ corresponds to the variable TECH1 (Access to a mobile phone—% age 15+) and so on (see Table 1 for the description of variables). The training set is a set of observations where each observation has one or more predictor variables $x_j$ and a single predicted variable $y$, called the output variable (or dependent variable).

The size of a training set is denoted by $m$. The set of output variables will be represented by a vector of size $m$.

**Table 1** Data description

| Variable name | Variable category | Mnemonics |
|---|---|---|
| Headcount of people living below the poverty line | DEP | POVERTY |
| Access to a mobile phone (% age 15+) | TECH | TECH1 |
| Access to internet (% age 15+) | TECH | TECH2 |
| Account (% age 15+) | FIN | FIN1 |
| ATMs per 100,000 adults | FIN | FIN2 |
| Agents of payment service providers per 100,000 adults | FIN | FIN3 |
| Borrowed from a financial institution or used a credit card (% age 15+) | FIN | FIN4 |
| Branches per 100,000 adults | FIN | FIN5 |
| Debit cards per 1000 adults | FINTECH | FINTECH1 |
| Deposit accounts per 1000 adults | FIN | FIN6 |
| E-money accounts per 1000 adults | FINTECH | FINTECH2 |
| Insurance policy holders per 1000 adults (life) | FIN | FIN7 |
| Insurance policy holders per 1000 adults (non-life) | FIN | FIN8 |
| Made or received digital payments in the past year (% age 15+) | FINTECH | FINTECH3 |
| Interoperability of ATM networks and interoperability of POS terminals (0–1) | FINTECH | FINTECH4 |
| Mobile agent outlets per 100,000 adults | FINTECH | FINTECH5 |
| Mobile money transactions per 100,000 adults | FINTECH | FINTECH6 |
| POS terminals per 100,000 adults | FINTECH | FINTECH7 |
| Retail cashless transactions per 1000 adults | FINTECH | FINTECH8 |
| Used a mobile phone or the internet to check account balance in the past year (% age 15+) | FINTECH | FINTECH9 |

Note: POVERTY is a dependent variable and all other variables are independent variables in the training data set

$$
Y = \begin{bmatrix} y_1 \\ y_2 \\ \dots \\ y_m \end{bmatrix}
$$

This vector denoted by $Y$ is called a feature vector and is in fact a matrix of dimension $m \times 1$. In our model, this corresponds to the independent variable—POVERTY (Headcount of people living below the poverty line). The goal of machine learning is to determine a function between a quantitative variable Y (output or dependent variable) and $n$ different predictor variables $x_1, x_2, \dots, x_n$ (input or independent variable). It is assumed that there is a relationship between $Y$ and $X = (x_1, x_2, \dots, x_n)$ determined by a function:

$$y = f(X) + \varepsilon.$$

In this case, $f$ represents a fixed but unknown function linking $(x_1, x_2, \ldots, x_n)$ to Y and including an error term $\varepsilon$ defined as a random distribution, independent of $X$ with a mean ($\mu = 0$). The model was tested with the training data and randomly cross-validated for the efficiency and performance of the model. The final model generated by the ML approach described above is described below:

$$\text{POVERTY} = C + \beta_1 \text{ TECH1} + \beta_2 \text{TECH2} - \beta_3 \text{FIN2} - \beta_4 \text{FIN4} + \beta_5 \text{FIN5}$$
$$- \beta_6 \text{FINTECH5} - \beta_7 \text{FINTECH6} + \beta_8 \text{FINTECH9} + \varepsilon.$$

## 4 Results and Analysis

The machine learning approach employed in this study is based on an iteration process developed on the correlation among the variables. In the initial iteration process, some of the variables were dropped due to high correlations, as that could have led to a spurious predictive model. The Pearson correlation measures the linear relationship between continuous variables, we measured the linear dependence between all variables included in the cleaned training data which is presented in Fig. 2 below. Generally, the correlation coefficient is between $-1$ and $+1$ and represents a scaled version of covariance providing the direction and strength of a relationship.

From the correlation analysis presented in Fig. 2 above, it is clear that almost all of the Technology (TECH), Conventional Financial Inclusion (FIN), and Fintech

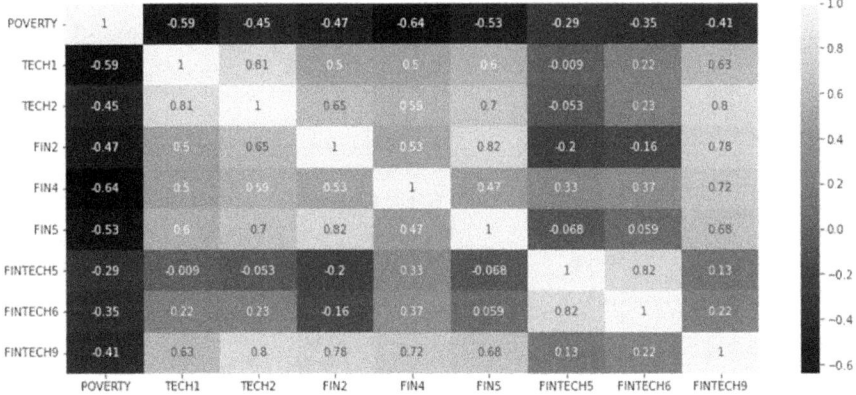

**Fig. 2** Correlation analysis

(FINTECH) variables are negatively associated with Poverty, thereby indicating that it might be useful to explore the relationship between technological penetration, financial inclusion, and fintech on poverty in the region and to predict a model that best presents the relationship between them. To that effect, we employed a machine learning (ML) predictive algorithm to develop a predictive model for our data, and for this study as earlier described the supervised learning algorithm predicted the multiple regression model described in the research methodology section. The representation of a validated model is presented below and the detailed results are presented in Table 2 below.

$$POVERTY = 84.98 - 0.56 \text{ TECH1} + 0.29 \text{ TECH2} - 0.85 \text{ FIN2} - 1.85 \text{ FIN4}$$
$$- 0.43 \text{ FIN5} - 0.016 \text{ FINTECH5} - 0.0000029 \text{ FINTECH6} + 1.20 \text{ FINTECH9}$$
$$+ \varepsilon.$$

The results presented in Table 2 suggest that the technological penetration, fintech development, and financial inclusion have a considerable impact on poverty alleviation in the African economies, except for the variables—"Access to Internet," and "Branches per 1000 adults." The results are consistent with the evolution and development of the financial inclusion landscape in Africa. These countries have

**Table 2** Regression results

| Dep. variable: | POVERTY | R-squared: | | 0.749 |
|---|---|---|---|---|
| Model: | OLS | Adj. R-squared: | | 0.643 |
| Log-likelihood: | −93.771 | F-statistic: | | 7.078 |
| AIC: | 205.5 | Prob (F-statistic): | | 0.0002 |
| BIC: | 217.5 | | | |
| | Coeff | Std err | t | P > |t| |
| C | 84.9844 | 8.305 | 10.233 | 0.000 |
| TECH1 | −0.5561 | 0.181 | −3.076 | 0.006 |
| TECH2 | 0.2969 | 0.351 | 0.847 | 0.408 |
| FIN2 | −0.8536 | 0.38 | −2.244 | 0.037 |
| FIN4 | −1.8526 | 0.621 | −2.2985 | 0.008 |
| FIN5 | −0.4278 | 1.096 | −0.39 | 0.701 |
| FINTECH5 | −0.0195 | 0.009 | −2.1667 | 0.033 |
| FINTECH6 | −3.03E-06 | 1.35E-06 | −2.2430 | 0.029 |
| FINTECH9 | 1.2028 | 0.436 | 2.757 | 0.013 |
| Omnibus: | 0.206 | Durbin Watson: | | 1.906 |
| Prob (omnibus): | 0.902 | Jarque-Bera (JB): | | 0.124 |
| Skew: | −0.139 | Prob (JB): | | 0.940 |
| Kurtosis | 2.828 | | | |

Notes: (1) Standard errors assume that the covariance matrix of the errors is correctly specified. (2) For the description of variables, see Table 1 pioneered a concept of highly successful mobile money (M-Pesa) and do not have sufficient brick-and-mortar infrastructure to operate financial operations in far-flung areas with inadequate road and internet infrastructure, and therefore, are heavily reliant on SIM-card-operated mobile money approaches to financial inclusion. The alternative disruptive models to financial inclusion have played a great role in expanding the outreach of financial services in the African region breaking beyond the conventional bank-operated financial inclusion frontiers. According to the World Bank Group (2019), in 2017, 43% of adults have an account with a bank or a mobile money agent relative to 34% of adults in 2014. Between 2014 and 2017, while a 4% increase was witnessed in the share of accounts held with a financial institution, the corresponding share of mobile money accounts almost doubled during the same period. The results indicate that among the technological penetration variables, access to mobile phones has a significant impact on poverty alleviation while access to the internet does not essentially help in reducing poverty. Among the conventional financial inclusion variables, ATM coverage and access to credit significantly contribute to reducing poverty while the brick-and-mortar outreach of banking infrastructure does not significantly reduce poverty. Among the fintech variables, breadth of mobile money agents, mobile money transactions, and use of mobile and internet to access accounts significantly contribute to the reduction of poverty in the African region while other fintech variables do not feature in the predictive model. These results are consistent with Tay et al., 2022, Kelikume (2021), Mohammed et al. (2017), Park and Mercado (2018), Khaki (2017), and Mushtaq and Bruneau (2019). The results suggest that expanding the outreach of mobile phone services has a positive impact on poverty alleviation which may be explained by the fact that access to mobile phones is the only way to access the digital financial inclusion platforms like M-Pesa in most African countries. Having a mobile phone essentially empowers people to access transaction services that can be used to operate the small farm and non-farm businesses, therefore, reducing vulnerabilities and helping the enterprise growth. Likewise, the results indicate that the conventional financial inclusion approaches like the breadth and depth of formal banking services like ATM Coverage and access to formal credit also act as an enabling factor in financial inclusion and significantly contribute to the economic dynamism, resilience, reducing vulnerabilities, and promoting economic growth, subsequently reducing poverty. Moreover, the results indicate that the fintech approaches to digital financial inclusion like the access to mobile money agents, the volume of mobile money transactions, and the use of mobile and internet to access accounts also act as enablers in the socio-economic development of otherwise excluded and poor segments and lifting them out of poverty. There is some evidence, though limited, consistent with the above findings available in the existing literature that indicate the synergies between conventional financial inclusion approaches and the fintech approaches to digital financial inclusion in poverty alleviation across different regions of the world (Appiah-Otoo & Song, 2021; Ye et al., 2022).

## 5   Conclusion

Financial inclusion has been a consistent policy goal for almost all the economies around the globe, particularly, those with less developed financial markets and institutions. Since the emergence of financial inclusion as a priority policy framework, the goal has been strategically pursued through different approaches, like priority sector lending, SHG lending, the Grameen approach, NGO-supported frameworks, brick-and-mortar approaches, the banking facilitator model, and so forth. The conventional approaches to financial inclusion have focused little on financial innovation and rather paid much attention mostly to the outreach mechanism. The conventional approaches to financial inclusion have recently been challenged and disrupted by innovative technology-oriented financial solutions collectively known as fintech. The potential of fintech has been found to be transformative for underdeveloped economies with poor financial infrastructure and has considerably disrupted the financial inclusion landscape across the globe. The pioneering advance in digital financial inclusion has been witnessed in the African region through the introduction of M-Pesa, a mobile money account that can be used to perform transactions just over a simple mobile phone connection. The growth of digital financial inclusion is believed to have reformative implications on the micro, small, and medium enterprises in the otherwise excluded regions and foster growth in economic activity, farm productivity, enterprise growth, employment, and subsequently, poverty alleviation and household welfare. The results of this study indicate that the mobile phone penetration, depth of financial services like ATM coverage and access to formal credit, the outreach of mobile money agents, the volume of mobile money transactions, and usage of mobile money and institutional accounts have a considerable impact on reducing poverty in African countries.

In view of the results of this study, some lessons and future directions for policy and regulation frameworks can be developed and revisited. Primarily, the national and regional policy frameworks on financial inclusion should build synergies among the conventional and emerging fintech solutions to financial inclusion and allocate resources for investment in fintech and the supporting infrastructure like mobile telephony and internet to (1) expand the outreach of financial services, and (2) increase the usage, efficiency, utility, and contribution of these financial inclusion platforms to foster micro, small, and medium enterprise development, saving and credit mechanisms, and economic growth.

On the regulation side, while the fintech landscape is expanding at a fast pace, it challenges the existing traditional mechanisms of financial inclusion motivating the traditional financial institutions to either lower the cost of their products and services or to expand their outreach. Having said that, the fintech platforms offer a serious regulatory challenge as they may extend beyond borders, and their operations may often be difficult to classify, monitor, and regulate. It is, therefore, important for the regulators to prepare a comprehensive regulatory policy that incorporates the integration of fintech and conventional approaches in a seamless way.

# References

Aggarwal, S., Brailovskaya, V., & Robinson, J. (2020, May). Cashing in (and out): Experimental evidence on the effects of mobile money in Malawi. In *AEA papers and proceedings* (Vol. 110, pp. 599–604).

Allen, F., Carletti, E., Cull, R., Qian, J., Senbet, L., & Valenzuela, P. (Eds.). (2020). *Improving access to banking: Evidence from Kenya*. Working paper.

Alliance for Financial Inclusion (AFI). (2013). Defining and measuring financial inclusion. In T. Triki & I. Faye (Eds.), *Financial inclusion in Africa*. African Development Bank.

Appiah-Otoo, I., & Song, N. (2021). The impact of fintech on poverty reduction: Evidence from China. *Sustainability, 13*(9), 5225.

Beck, T. (2020). *Fintech and financial inclusion: Opportunities and pitfalls*. ADBI working paper series, no. 1165. Asian Development Bank Institute (ADBI).

Beck, T., Pamuk, H., Ramrattan, R., & Uras, B. R. (2018). Payment instruments, finance and development. *Journal of Development Economics, 133*, 162–186.

Buchak, G., Matvos, G., Piskorski, T., & Seru, A. (2018). Fintech, regulatory arbitrage, and the rise of shadow banks. *Journal of Financial Economics, 130*, 453–692.

Churchill, S. A., & Marisetty, V. B. (2020). Financial inclusion and poverty: A tale of forty-five thousand households. *Applied Economics, 52*(16), 1777–1788.

Claessens, S., Frost, J., Turner, G., & Zhu, F. (2018, September). Fintech credit markets around the world: Size, drivers and policy issues. *BIS Quarterly Review*, 29–49.

Demirgüç-Kunt, A., & Klapper, L. F. (2012). *Financial inclusion in Africa: An overview*. World Bank policy research working paper (6088).

Demirgüç-Kunt, A., Klapper, L. F., Singer, D., & Van Oudheusden, P. (2015). *The global findex database 2014: Measuring financial inclusion around the world*. Policy research working paper 7255. World Bank.

Erlando, A., Riyanto, F. D., & Masakazu, S. (2020). Financial inclusion, economic growth, and poverty alleviation: Evidence from eastern Indonesia. *Heliyon, 6*(10), e05235.

Faye, I., & Triki, T. (2013). Financial inclusion in Africa: The transformative role of technology. In T. Triki & I. Faye (Eds.), *Financial inclusion in Africa*. African Development Bank.

Fuster, A., Plosser, M., Schnabel, P., & Vickery, J. (2019). The role of technology in mortgage lending. *Review of Financial Studies, 32*, 1854–1899.

GPFI. (2016). Global partnership for financial inclusion: China 2016 priorities paper. G20, China 2016. Retrieved June 29, 2022, from https://www.gpfi.org/publications/global-partnership-financial-inclusion-gpfi-china-2016-priorities-paper

Hotchkiss, G., & Chuen, D. L. K. (2018). From the ground up: The financial inclusion frontier. In *Handbook of blockchain, digital finance, and inclusion* (Vol. 2, pp. 405–429). Academic Press.

Inoue, T. (2018). Financial inclusion and poverty reduction in India. *Journal of Financial Economic Policy, 11*, 31.

Kelikume, I. (2021). Digital financial inclusion, informal economy, and poverty reduction in Africa. *Journal of Enterprising Communities: People and Places in the Global Economy, 15*, 626–640.

Khaki, A. R. (2017). Does access to finance alleviate poverty? A case study of SGSY beneficiaries in Kashmir Valley. *International Journal of Social Economics, 44*(8), 1032–1045.

Khaki, A., & Sangmi, M. U. D. (2016). Financial inclusion & social capital: A case study of SGSY beneficiaries in Kashmir Valley. *Independent Journal of Management AND Production, 7*(4), 1005–1033.

Kikulwe, E. M., Fischer, E., & Qaim, M. (2013). *Mobile money, market transactions, and household income in rural Kenya* (No. 22). Global food discussion papers.

Lauer, K., & Lyman, T. (2015, March). *Digital financial inclusion*. Consultative Group on Assisting the Poor (CGAP).

Mbiti, I., & Weil, D. N. (2015). Mobile banking: The impact of M-Pesa in Kenya. In *African successes, Volume III: Modernization and development* (pp. 247–293). University of Chicago Press.

Mehrotra, A. (2019). Financial inclusion through fintech–a case of lost focus. In *2019 International Conference on Automation, Computational and Technology Management (ICACTM)* (pp. 103–107). IEEE.

Mohammed, J. I., Mensah, L., & Gyeke-Dako, A. (2017). Financial inclusion and poverty reduction in sub-Saharan Africa. *African Finance Journal, 19*(1), 1–22.

Mushtaq, R., & Bruneau, C. (2019). *Microfinance, financial inclusion and ICT: Implications for poverty and inequality.* Technology in Society.

Park, C. Y., & Mercado, R., Jr. (2018). Financial inclusion, poverty, and income inequality. *The Singapore Economic Review, 63*(01), 185–206.

Philippon, T. (2019). *On fintech and financial inclusion* (No. w26330). National Bureau of Economic Research.

Sapovadia, V. (2018). Financial inclusion, digital currency, and mobile technology. In *Handbook of blockchain, digital finance, and inclusion* (Vol. 2, pp. 361–385). Academic Press.

Soyemi, K., Olowofela, O., & Yunusa, L. (2020). Financial inclusion and sustainable development in Nigeria. *Journal of Economics and Management, 39*(1), 105–131.

Stein, P., Bilandzic, N., & Hommes, M. (2013). Fostering financing for Africa's small and medium enterprises. In T. Triki & I. Faye (Eds.), *Financial inclusion in Africa.* African Development Bank.

Suri, T., & Jack, W. (2016). The long-run poverty and gender impacts of mobile money. *Science, 354*(6317), 1288–1292.

Tay, L. Y., Tai, H. T., & Tan, G. S. (2022). Digital financial inclusion: A gateway to sustainable development. *Heliyon, 8,* e09766.

Wang, X., & Fu, Y. (2021). Digital financial inclusion and vulnerability to poverty: Evidence from Chinese rural households. *China Agricultural Economic Review, 14,* 64–83.

Wieser, C., Bruhn, M., Kinzinger, J., Ruckteschler, C., & Heitmann, S. (2019). *The impact of mobile money on poor rural households: Experimental evidence from Uganda.* The World Bank.

World Bank. (2020a). How regulators respond to fintech; Evaluating the different approaches – Sandboxes and beyond. In *Finance, competitiveness & innovation; global practice.* Fintech note, no. 5. World Bank Group.

World Bank. (2020b). Digital financial inclusion. Retrieved June 28, 2022, from https://www.worldbank.org/en/topic/financialinclusion/publication/digital-financial-inclusion

World Bank Group. (2019). Global Financial Inclusion and Consumer Protection Survey, 2017 Report. World Bank, Washington, DC. © World Bank. https://openknowledge.worldbank.org/handle/10986/28998. License: CC BY 3.0 IGO.

Ye, Y., Chen, S., & Li, C. (2022). Financial technology as a driver of poverty alleviation in China: Evidence from an innovative regression approach. *Journal of Innovation and Knowledge, 7*(1), 100164.

Milton Keynes UK
Ingram Content Group UK Ltd.
UKHW020740141123
432548UK00007B/475